rororo sprachen

Herausgegeben von Ludwig Moos

The Way Things Work vermittelt schnell, effektiv und mit Witz jenes spezielle Englisch, das uns verstehen läßt, wie Dinge funktionieren. Ob Sie wissen müssen, wie man über die Produkte spricht, mit denen Ihre Firma zu tun hat, oder englische Beschreibungen aus Privatinteresse verstehen möchten, hier finden Sie den grundlegenden technischen Wortschatz und die zugehörigen Redewendungen. Grafiken unterstützen das Verständnis, die Dialoge sparen Pleiten und Pannen mit der Technik nicht aus.

Dr. René Bosewitz ist Native Speaker und unterrichtet an der Berufsakademie Mannheim. In einer deutschen Zweigstelle der London Chamber of Commerce bereitet er Firmenangehörige auf einschlägige sprachliche Prüfungen vor. Er trainiert zudem seit vielen Jahren Manager aus Banken und Industrie in Business English. Mit Robert Kleinschroth hat er bei rororo sprachen *Joke Your Way Through English Grammar* (8572) und *Joke by Joke to Conversation* (8795) veröffentlicht.

Robert Kleinschroth unterrichtet Englisch am Gymnasium und an der Universität Heidelberg. Er hat zwanzig Jahre Praxis in der Erwachsenenbildung und leitete fünfzehn Jahre lang die Sprachabteilung eines Großunternehmens. Robert Kleinschroth hat zusammen mit Dieter Maupai *La Grammaire en s'amusant* (8714) verfaßt und außerdem *La Conversation en s'amusant* (8873) sowie *Sprachen lernen* (9140) und *Neues Lernen mit dem Computer* (9841) geschrieben.

In der Reihe Business English sind von beiden Autoren außerdem erschienen: *Manage in English* (60137), *Better than the Boss* (60138), *How to Phone Effectively* (60139), *Drop Them a Line* (60261, mit Bryan Hemming), *Test Your Management Skills* (60260), *Get Through at Meetings* (60262), *Let's Go International* (60267), *Check Your Language Level* (60268) und *Business English from A to Z* (60269, mit Bryan Hemming).

RENÉ BOSEWITZ/ROBERT KLEINSCHROTH

THE WAY THINGS WORK

TECHNISCHES ENGLISCH

FÜR BUSINESS UND ALLTAG

Rowohlt

rororo

Wir danken Frau Ortrud Grimm vielmals für ihre Hilfe
bei der Erstellung des Manuskripts.
Ohne sie hätten wir es wohl nicht geschafft.

Originalausgabe
Veröffentlicht im
Rowohlt Taschenbuch Verlag GmbH,
Reinbek bei Hamburg, Mai 1998

Umschlaggestaltung Barbara Hanke
(Illustrationen Cathrin Günther)
Layout Anne Drude und Alexander Urban
Satz OCRA und Stone PostScript,
QuarkXPress 3.32 (Dolev 800)
Druck und Bindung Clausen & Bosse, Leck
Printed in Germany
ISBN 3 499 60369 1

Inhalt

Guten Tag. Ich bin Ihr Buch. Vielen Dank, daß Sie mich aufgeschlagen haben – Sie müssen ein neugieriger und bildungsbewußter Mensch sein. Leisten Sie mir doch noch für ein paar Zeilen Gesellschaft. Wollen Sie lernen, wie die Dinge auf englisch funktionieren, die Ihnen auf Schritt und Tritt, in der Firma, zu Hause und im Ausland begegnen? Dann schließen Sie mit mir eine Lücke. Mein Methode und mein Inhalt:

- learning by laughing,
- everyday technical English,
- a business English background

Nach einem Spaziergang durch die Abteilungen der traditionsreichen Firma Ruver Cars, Ltd. können Sie im englischsprachigen Ausland mitreden: über Bürokommunikation, Computer, Fotokopierer und vieles mehr. Aber auch die Welten der kleinen, nützlichen Dinge eröffnen sich Ihnen. Oder wissen Sie etwa schon, was da alles auf Ihrem Schreibtisch steht – der Locher, die Heftmaschine oder das namenlose Ding, mit dem man die Heftklammern aus Papieren wieder entfernt? Auch wenn Sie beruflich nichts mit Motoren, Computern, Druckern, Scannern, Telefonen, Kopierern oder Ziegelsteinen zu tun haben, werden Sie feststellen, daß ein Großteil des Vokabulars der Alltagssprache entnommen ist oder sich auf Ihr Gebiet übertragen läßt. Und noch etwas: Die Unterhaltung soll auch nicht zu kurz kommen. Alles ist mit viel englischem Witz gewürzt.

Time for a smile

Dr. Richard Block ordered a go-cart for his son Jim's birthday. It arrived packed in a large box which contained[1] 123 pieces. The instructions said that it could be assembled[2] in one hour. After a week, just in time for Jim's birthday he had managed to fit the pieces together.
Dr. Block wrote a check, tore[3] it into 123 pieces, and sent it to the supplier.

1 enthalten 2 zusammenbauen 3 tear, tore, torn: zerreißen 4 Lieferfirma

Gebrauchsanweisung

Alle Kapitel sind in eine lockere Rahmenhandlung eingebunden. Jedes Kapitel erzählt eine Episode aus dem Leben des amerikanischen Rauhbeins James Kelly. Sie kennen ihn vielleicht schon aus anderen Bänden unserer Business English-Reihe. Diesmal zeigt er den Engländern, wie amerikanischer Pioniergeist und Pragmatismus Probleme lösen. Davon gibt es genug, denn Ruwer wird zum Leidwesen der Belegschaft erst von Gondex, Japan, und später von AMW (Auto Motorenwerke, Germany) übernommen. Mit der neuen Firmenphilosophie hält jedesmal die Technologie des jeweiligen Mutterlandes Einzug in Büros und Werkstätten.

- **Our story:** Die Episoden unserer Geschichte bilden die *storyline* für die abwechslungsreichen Übungsmöglichkeiten. Dabei lassen wir Ihnen sehr viel *Time for a smile* oder geben Ihnen *Food for thought,* Witze und Zitate aus den spitzen Federn großer Männer und Frauen. Sie bringen nicht nur Sonne in das Tal des Lernens, sondern beleuchten das Thema von einer höheren Warte.
- **Working with the text:** Sie testen Ihr Textverständnis und erweitern Ihren Sprachschatz.
- **How things work in English** ist das Kernstück jedes Kapitels. Sie lernen und testen das technische Englisch für den gebildeten Laien und interessierten Geschäftsmann.
- **Situations:** Testen und erweitern Sie Ihre *speech functions,* die Sie für Geschäftssituationen wie Besprechungen oder Reklamationen brauchen.
- **Beef up your word power:** Kleine gehirngerechte Lernhappen vertiefen, was Sie von den vorangegangenen Stationen mitgenommen haben.
- **Our quiz for the experts:** Einige Kapitel laden zu einem Quiz ein, das auch Sie mit Ihrer Schulweisheit lösen können.
- **Extra reading:** Belohnt Sie mit Verrücktem, Absurdem oder Groteskem zum Thema. Solch «merkwürdige» Texte, werden Sie garantiert vergeblich in den Schulbüchern oder Kursmaterialien seriöser Autoren suchen.

- **Let's do it crosswise:** Für unsere Rätselfreunde das *crossword puzzle* zum Thema, das den Wortschatz des Kapitels wiederholt.
- **Vocabulary:** Im Anschluß an jedes Kapitel finden Sie eine ausführliche Vokabelliste.
- **Key:** Im Anhang stehen die Lösungen zu den Übungen.

Pathfinder to situational English

And now Ladies and Gentlemen, proceed to Chapter 1. James Kelly is waiting for you.

How it all began

Technical English? In business? Well, that's a tall order, but we think we've got the answer! We'll give you the basic stuff for our everyday man in business, in the company, in the "real" world. It's one of our heroes who tells us about it and learns the hard way from experience as he is confronted by various technical situations. And in those situations you, the reader, can pick up some of the terminology associated with technical things in the business world.

Who's the hero this time? Remember Jamie in "Better than the Boss", remember Jamie in "Let's Go International"? Well, it's the type again. It's James. James Kelly. We're sure you met the type somewhere. We all have. He's a big, solid American from Ohio. He's pushed his way through, got qualified, got an "education". Solid stuff. About 40 years of age, slow-talking, deep voice. At the age of 17 he got apprenticed to a big lumberyard in Ohio that sold wood to home centres. However, he soon realised that wood is one of the few materials that will rot as well as burn and that if God had wanted him to sell it, he wouldn't have made plastic and aluminium. He imagined that there were more challenging products to market. So he cut his teeth as a salesman on the international market for SASL, a manufacturer of intelligent Sports Aids for the Slow Learner.[1] There he became interested in technical stuff, sold photocopiers for Law Valley Copiers, in Germany[2], and then got a thing about cars. He even joined General Motors for a few months and then did some courses in technology for the automobile industry.

One day he saw this advertisement in the Times for a technical manager. The company advertising was Ruver, a leading car manufacturer in Britain.

1 Read all about it in *Let's Go International.* Business English rund um die Welt (rororo sprachen).
2 You might have met James in *Better than the Boss.* Business English fürs Büro(rororo sprachen).

Ruver Cars, Ltd.

We are a leading group in Great Britain, we specialise in manufacturing ranges of quality cars.

Our name speaks for itself.
We require: one senior technical manager.

Duties (There are many):
➨ To support various departments at the headquarters in England, ensure that systems function optimally, to enhance processes at all levels.

Qualifications (There should be many):
➨ Five years experience with similar responsibility. All appropriate academic qualifications.

● **Salary package: £ 90,000 plus extras.**

Please contact:
R. Bossybitch, Executive Search,
London SW · 32, Wilton Crescent

Well, Jamie got the job. They even arranged his work permit. Let's see how James describes his experience.

Time for a smile: Unusual job advertisements

1. Window cleaner to clean the windows on envelopes.
2. Proof-reader for a skywriting company.
3. Translator for bad mimes.
4. One-legged man for an arse-kicking competition.
5. Lifeguard in a car wash

Vocabulary

ad; advert; advertisement	Anzeige, Annonce
aid	Hilfe
appropriate	angemessen
arse-kicking competition	Wettbewerb im Hinterntreten
car manufacturer	Autohersteller
car wash	Autowaschanlage
cut one's teeth	Erfahrungen sammeln
duties	Pflichten
enhance	verbessern
ensure	sicherstellen, gewährleisten
envelope	Briefumschlag
experience	Erfahrung
hero	Held
leading	führend
level	Ebene
lifeguard	Rettungsschwimmer
lumberyard	Holzgroßhandel
manufacturer	Produzent, Hersteller
mime	Pantomime
one-legged man	Einbeiniger
pick up	aufsammeln
proof reader	Korrektor
push one's way through	seinen Weg machen
range	Baureihe, Palette
require	benötigen
rot	verfaulen
similar	ähnlich
responsibility	Verantwortung
tall order	schwere Aufgabe
various	verschiedene
work permit	Arbeitserlaubnis

Motor-car misery

Time for a smile: In a public telephone box

She: Hello, darling. Is that you? I'm sorry to trouble you, but we've got a little problem with the new car. I ...

He: The new car? What's wrong with it, darling?

She: I don't know how to explain this, darling, but the motor won't start. There seems to be some water in the carburettor.

He: Don't cry, darling. I'll be with you in a moment. By the way, where is the car now?

She: Well, darling, ... it is in the River Thames, near Greenwich.

(*) Breakdowns and burning bottoms

James' diary

Hi folks, I'm Jamie. I'm going to guide you through my business experiences of the last two years. You know I was hanging around the car business doing this and that for General Motors, and I did a bit of studying, too. Then I saw this ad for Europe. A chance to go to Britain. My field! They wanted somebody to check technical systems at the headquarters. I gave it my best shot. Hit the bullseye. And wowie! Here I am. And that's where it all began.
After my plane landed in London I had to get to the Ruver head quarters. They sent me a brand new Ruver taxi. I can tell you I was impressed. Here's what happened on the way.

James (**J**), *company chauffeur* (**C**)

J: Wow, I've been all over the place, but this is the first time I've actually had a job in England. It's great to be here. Which state are you from?

C: Me, I'm not from a state. I'm from Nottingham, but I've been living down here in London for eight years. Came in 1989.
(noise of grating gears)

J: Jesus man, be careful, what was that?

C: Something not properly synchronised in the gear-box.

J: No automatic box?

C: Not on this model. And if there were, I guess it wouldn't make much difference. In my opinion these engineers can't read specifications.

J: Hey, man, you're shooting at my new company. Don't you like your job?

C: I like my job. It's income. And I'm into cars.

J: Especially Ruver. It's great. Bet you're proud to be part of this set-up?

C: I don't know about proud, but the product is certainly better than some continental cars. Just remember the case of Polec. Their cars tended to catch fire at the petrol stations. And those big taxi-like cars from Muckadiz; they were dragging along the road when there were three passengers in the car and ...

J: *(breaking in)* Okay, okay! But anyway this is a Ruver and I'm sure there's nothing much wrong with it.

C: Hmm. We'll see. We have to drop in for a check at one of our garages on the way. I hope you don't mind. We've got electronics in the car.

J: I know we have electronics. You know every modern car ...

C: I mean they don't function properly. The chip's up the spout.

J: Spout? What does it all have to do with tea?

C: Broken, defective, out of service.

J: I'm sure that's just some minor detail.

C: Listen mate! I'll give you some examples of minor details. We've got to turn right at the lights. I'll signal.

J: That's normal!

C: Wait for it!

J: Ouch! My bottom's burning. Please switch off the seat-warmer.

C: Actually it was just the indicator. I'll just switch on the windscreen wipers.

J: My god, the sunroof has opened. And it's raining!

C: Well, that'll cool your burning bottom. That's technology for you. But take it easy. They've got a diagram in the garage.

J: Of the electronics?

C: No, just the normal mechanics. Just the ordinary parts. Here we are! *(Fifteen minutes later)*

Mechanic: So here's the report. Show this to your maintenance department, and the R&D for that matter. Here are the main parts of the car. Defective components are marked with an X. There are quite a few things.

J: Wow! I thought this was a new car! Problems like these are normally a sign of age, you know, built-in obsolescence. We're good at that in the States, too.

Mechanic: Sorry to disappoint you, mate. It's the latest model, a prototype. They test it on the staff, we're all guinea pigs at the company. The electronics, well, they need a re-design. It's still full of teething problems.

James' diary

Well folks, you can imagine what I felt with all that happening on my first day in the new company. But after all, that's what I'm here for. I went straight to the technical director.

T Working with the text

1. Are the following statements true (T) or false (F)?

1. James is in England for the first time. ☐
2. The company taxi has an automatic gear-box. ☐
3. There's something wrong with the electronic system. ☐
4. The taxi is a test car. ☐
5. The defects are repaired at the garage. ☐

2. What they might have said but didn't

Replace the words in italics with those used in the dialogue.

1. The engineers can't read the *instructions.* ....
2. We have to *stop* at one of our garages. ....
3. I'll just *turn on* the windscreen wipers. ....
4. Defective *parts* are marked with an X. ....
5. The car's still full of *preliminary* problems. ....
6. Ruver cars are *the best.* ....

Food for thought

The guy who invented the first wheel was an idiot, but the guy who invented the other three, now he was a genius.

Sid Caesar

H How things work in the car

Toolbox 1: For safety's sake

horn	Hupe	indicator	Blinker
headlights	Scheinwerfer	steering wheel	Steuerrad
rear-view mirror	Rückspiegel	windscreen wiper	Scheibenwischer

and don't forget to fasten your seat belt!

1. Tools in use: What to do if you want to ...

show a new direction	....	the indicator
warn a pedestrian	....	the horn
avoid an obstacle	....	the steering wheel
clean the windshield /- screen	....	the windscreen wipers
drive at night	....	the headlights
see what's going on behind	....	the rear-view mirror

Toolbox 2: Pedals for your feet (from right to left)

accelerator	Gaspedal
brake	Bremse
clutch	Kupplung

2. Tools in use: Complete!

you want to stop the car	put the	.... on
you want to go faster	step on the	....
you want to go into neutral	push in the	....

Situations

1. In case of breakdowns – don't worry, be happy

Typical problems ...	and what to say at the garage
1 you've run out of petrol	A put on / fit the spare
2 there isn't enough oil	B make it watertight / seal it
3 the clutch is slipping	C recharge it / charge it up
4 you have a puncture	D try to bump-start it
5 the battery is flat	E adjust it
6 the radiator is leaking	F tighten it
7 the car won't start	G weld the hole in exhaust pipe
8 the fanbelt is slipping	H try to seal the sump
9 the motor is too loud	I top her / it up
10 the oil is leaking	J adjust the ignition / check the plugs
11 the engine sounds rough	K fill her / it up

Be polite and add 'please' each time you ask for help.

2. Wrong order: a driving test

Get into the car and do it – but in the right order.
The first two have been done for you

- [ ] Push in the clutch.
- [ ] Press the accelerator, drive away.
- [ ] Release the handbrake.
- [ ] Put the car into first gear.
- [ ] Push the clutch in again.
- [ ] Release the clutch very slowly.
- [2] Turn it and start the engine.
- [ ] Let the clutch out.
- [1] Put the key into the ignition.
- [ ] Shift / change gear and so on.

Well done! However, you're driving
on the wrong side of the road. We're in Britain.

3. Do it yourself!

Bad luck! You've got a puncture. What will you do? Let's take a look at the handbook. Typical! Must have been written by a lunatic or a computer. Before you start, number the steps in chronological order.

☐ Open the boot.
☐ Jack up the car.
☐ Detach the hubcap with a screwdriver.
☐ Open the tool kit and take out the wheelbrace and the jack.
1 Apply the handbrake.
☐ Loosen the wheelnuts, but don't unscrew them completely.
☐ Put on the spare.
☐ Take the wheelbrace and tighten the screws
☐ Stop at the next petrol station and check the air pressure.
☐ Take out the spare wheel.
☐ Replace the hubcap.
☐ Unscrew the wheelnuts completely; remove the wheel.
☐ Replace the wheelnuts and hand-tighten them.
☐ Lower the car.

There you are!

W Beef up your word power

1. Odd man out – Which item doesn't fit into the list?

A	B	C	D
brake	tyre	radiator	antenna
pedal	hubcap	carburettor	glove compartment
clutch	wheelnuts	boot	dash board
accelerator	jack	battery	gear shift

2. In other words

Find a word that has (almost) the same meaning

component	....
switch on the headlights	....
let out the clutch	....
take off the hubcap	....
fill up the oil	....
change gear	....

3. Explain it with the opposite

lower the car	....	the car
switch on the headlights	....	the headlights
slow down the car	....	the car
release the clutch	....	the clutch
screw it on	....	it
discharge the battery	....	the battery
loosen the wheelnuts	....	the wheelnuts

Q Our quiz for the experts

Do you know the combustion engine?	Yes	No
It converts heat into motive power.	☐	☐
Jets and rockets are combustion engines, too.	☐	☐
Pistons rotate inside the cylinders.	☐	☐
Strong vehicles have two-stroke engines.	☐	☐
All combustion engines have spark plugs.	☐	☐
Diesels have inlet and outlet valves.	☐	☐
Four-stroke engines have four valves.	☐	☐
Combustion engines burn air and petrol.	☐	☐
74 kWh is about 100 horsepower.	☐	☐
Cars waste 60% of their energy.	☐	☐

Ⓡ Extra reading: A Car Called Horse

Interviewer (**I**); *Inventor Carl Perscho* (**P**)

I: Mr Perscho, at this year's automobile fair in Hannover you presented an interesting alternative to the off-road car.

P: You're referring to the latest version of my autohorse. It has all the advantages of the off-road car as well as the compact car.

I: What are the advantages of this most astonishing vehicle?

P: It can drive over stones, cross fields and forests. A car cannot. You see, the main handicap of a car is its wheels. The most natural and efficient way of moving forward is the leg.

I: I see. That's why you invented a vehicle on legs.

P: I didn't invent it. I just copied Mother Nature. Well, the engine sits inside the body and controls each of the four legs. The motor is started by a crankshaft where the horse's tail is normally located ...

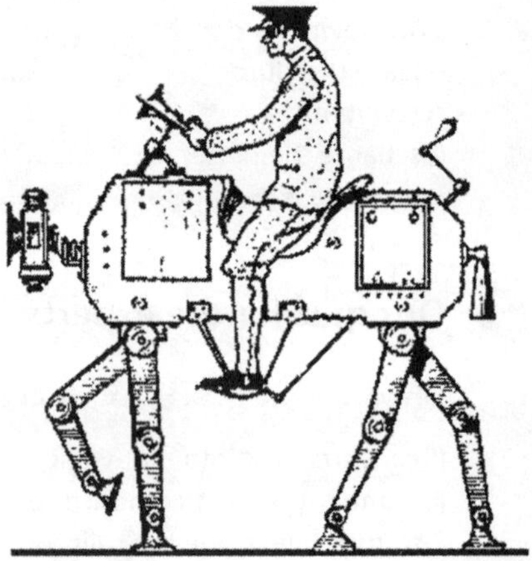

I: ... and that's where the exhaust gases escape, is it?

P: I see you've got the idea. Where else should they escape from? It's the most likely spot on the construction.

I: Great! Legs instead of wheels. It's so simple, so natural.

P: Well, it does have one wheel, after all. A steering wheel. That controls each of the legs separately. And my de luxe version has collapsible trays on each side of the body. The driver – or should I say the rider – can use them for having a picnic.

I: Thank-you, Mr Perscho for this fascinating interview.

(Adapted from the January 1905 issue of Automobilwelt)

C Let's do it crosswise

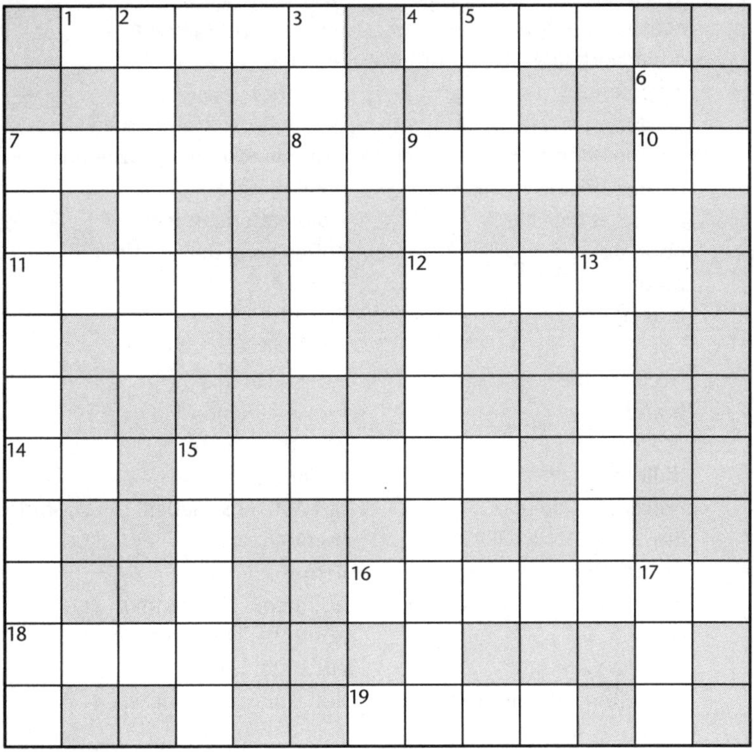

Across

1 the fifth wheel
4 you put your luggage there
7 another word for 'car'
8 noise in the gear-box
10 short for 'advertisement'
11 a jack is a – ? –
12 they do it at garages
14 run out ? petrol
15 defective
16 it supplies energy
18 oil and water run out of it
19 you – ? – holes in pipes

Down

2 combustion engines need it
3 motor
5 let the clutch – ? –
6 a tool in the boot
7 long form of 'auto'
9 opposite of 'slow down'
13 shift
15 the pedal in the middle
17 don't – ? – out of petrol

Ⓥ Vocabulary

accelerator	Gaspedal
activate the signal lever	den Blinker betätigen
ad; advert; advertisement	Anzeige
air pressure	Luftdruck
apply	betätigen; sich bewerben
~ the handbrake	die Handbremse anziehen
battery; recharge the ~	die Batterie aufladen
the ~ is flat	die Batterie ist leer
boot; open the ~	Kofferraum; den ~ öffnen
bottom	Hintern
brake	Bremse
step / put on the ~	auf die ~ treten
release the ~	die ~ lösen,
brand new	funkelnagelneu
broken	kaputt
built-in	eingebaut
bump-start the car	das Auto anschieben, um zu starten
burn	brennen
carburettor	Vergaser
catch fire	brennen, Feuer fangen
check	Kontrolle, Überprüfung
check a system	ein System prüfen
clutch; press / push in the ~	Kupplung; die ~ drücken
the ~ is slipping	die ~ rutscht
adjust the ~	die ~ nachstellen
let the ~ out	die ~ loslassen
combustion engine	Verbrennungsmotor
component	Bauteil
convert	umwandeln
cylinder	Zylinder
dashboard	Armaturenbrett
defective	fehlerhaft
detach	abnehmen, losmachen
direction; change ~	Fahrtrichtung ändern

drag along the roads	hier: Funken schlagen wegen mangelnder Bodenfreiheit
driving test	Fahrprüfung
drop in at a garage	bei einer Werkstatt vorbeischauen
engine; start the ~	Motor; den ~ anlassen
two-/four-stroke ~	Zweitakt-/Viertaktmotor
exhaust pipe	Auspuff
fanbelt; the ~ is slipping	Keilriemen; der ~ rutscht durch
fasten seatbelt	Sicherheitsgurt anlegen
fill her up, please	bitte volltanken
garage	Werkstatt
gear	Gang
gear lever	Schaltknüppel
change / shift gear	schalten
put into first ~	in den ersten ~ schalten
gear-box	Getriebe
gearshift	Gangschaltung
glove compartment	Handschuhfach
grate	knirschen, kratzen
guide; ~ someone through ...	führen; jemanden durch ... ~
guinea pig	Versuchskaninchen
handbrake	Handbremse
hang around	herumstehen, herumlungern
headlights	Scheinwerfer
horn	Hupe
horsepower	Pferdestärke
hubcap; detach the ~	Radkappe; die ~ abnehmen
ignition	Zündung
put the key into the ~	den Zündschlüssel einstecken
income	Einkommen
indicator	Blinker
instruction	Gebrauchsanweisung
jack	Wagenheber
jack up the car	den Wagen aufbocken
latest model	neuestes Modell
leak n., v.	Leck; undicht sein

lever	Hebel
lower the car	das Auto herunterlassen; abbocken
maintenance department	Kfz-Instandsetzung
mate	Kumpel
neutral; go into ~	den Gang rausnehmen; in den Leerlauf schalten;
obsolescence	künstliche Veralterung
obstacle; avoid an ~	einem Hindernis ausweichen
oil change	Ölwechsel
opinion; in my ~	meiner Meinung nach
part	Teil
pedal	Pedal
pedestrian	Fußgänger
piston	Kolben
plug	Zündkerze
preliminary problems	Anfangsprobleme
proud	stolz
puncture	Plattfuß
push; give the car a ~	das Auto anschieben
radiator	Kühler
the ~ is leaking	der ~ ist undicht
make the ~ watertight	... den ~ abdichten
rear-view mirror	Rückspiegel
re-design	neu konstruieren
replace the hubcap	wieder anbringen; die Radkappe ~
rotate	sich drehen
run out of petrol	kein Benzin mehr haben
safety	Sicherheit
sake; for safety's ~	der Sicherheit zuliebe
screwdriver	Schraubenzieher
seatbelt	Sicherheitsgurt
service; be out of service	nicht funktionieren
set-up	Organisation; hier: Firma
shot; give something one's best ~	sein Bestes geben
sign of age	Alterserscheinung
signal	hier: Blinker setzen

signal lever	Blinkhebel
silencer	Auspufftopf; Schalldämpfer
slip, the clutch is slipping	die Kupplung rutscht durch
sound the horn	hupen
spare (spare wheel)	Ersatzrad
put on the ~	das Ersatzrad montieren
spare part	Ersatzteil
spark plugs	Zündkerzen
specifications	Lastenheft; Spezifikation
spout; be up the ~	defekt, im Eimer sein
steering wheel	Steuerrad
sunroof	Schiebedach
switch on	einschalten
teething problems	Kinderkrankheiten
tend to	neigen zu
tighten	anziehen, straffen
tool kit	Werkzeugkasten
top up the oil	Öl nachfüllen
tops; be ~	Spitze sein
tyre	Reifen
valve	Ventil
inlet / outlet valve	Einlaß-/Auslaßventil
waste energy	verschwenden; Energie ~
weld	schweißen
wheelbrace	Kreuzschlüssel
wheelnut	Radmutter
loosen the ~	die ~ lösen
hand-tighten the ~	die ~ mit der Hand anziehen
unscrew the ~	die ~ abschrauben
windscreen	Windschutzscheibe
windscreen wiper	Scheibenwischer
windshield	Windschutzscheibe

3 Chapter

Car styling

Food for thought

US businessman Chauncey Depew warned his nephew not to invest $5,000 in an unknown young man named Henry Ford. He told him: "Nothing has come along that can beat the horse and the buggy."

(*) A matter of taste

James' diary

Hi folks. Well, I'm not so sure I'll be running to the director the next time there's a problem. He treated me as if I didn't know which way was up. He reckoned I knew all about the defects. "And that was the reason why we employed you, Mr Kelly, wasn't it?" I was pretty damned angry. And he sent me to the R & D styling department for the automobile interior. Well, I was only trying to help. Here's what happened.

In the design department: James (J), *Carl Bennet – chief designer* (C)

J: *(fade in)* ... and so the director sent me down to collect information about the present styling process. Perhaps a little bit of the American approach can help to get this thing off the ground.

C: *(angrily)* Well, I hardly think that American styling is the answer to our problems when I look at the cars you've produced in the last ten years. Gas-guzzling, heavyweight tanks more suited to the battlefield than the motorway.

J: You mean 'freeway'. Yes, I admit that most American models are more suited to wide open spaces, but we've refined our designs in the last five or six years. Everything is smaller, more compact. And our two major companies are the most successful all over the world. We've bought a lot of your English car industry.

C: But not Ruver. We're still independent.

J: For how long, gents? For how long? Come on. Don't take it to heart. I'm on your side. That's why I'm here. Show me a general overview.

C: Okay, here's a view of the front interior, the dashboard. But the main problem's with the electronics.

J: I've experienced the electronics first hand or rather first bottom. You should let others worry about electronics. Your problem's staring you in the face. Look at the diagram of the latest four-door model. It's unbelievable!

C: Well, we thought so, too. Still, it doesn't seem to be selling.

J: No man, you're not getting my message. It must be the language. You British might say it's bloody awful.

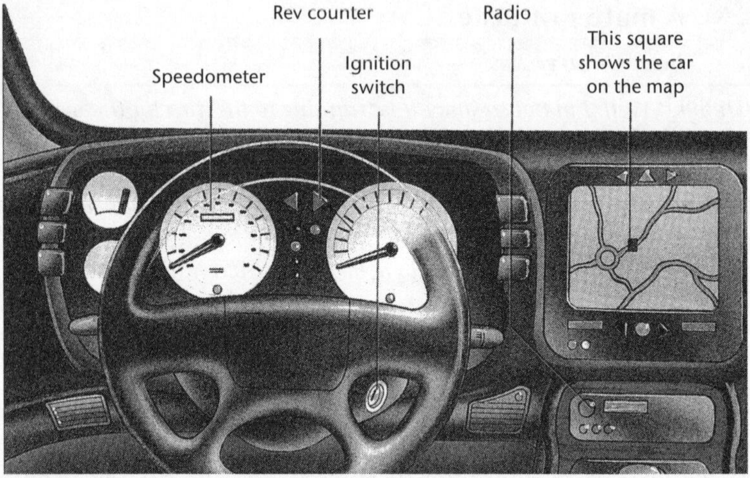

C: I say, I say. There's no need to swear. Perhaps there could be a few minor alterations but ...

J: *(interrupting)* I wouldn't even use it for a taxi-cab. Look at it! You English like to put crosses everywhere. I'll put a cross against each out-of-date feature *(scratching of a pencil)*. There you are!

C: Hey, hang on a minute! What's wrong with the speedometer? That's a must for every modern car.

J: Of course it is. A modern speedo for a modern car! Look at this old, circular design. Scrap it! You need digital information. It should be in the form of a bar graph. Here, like this.

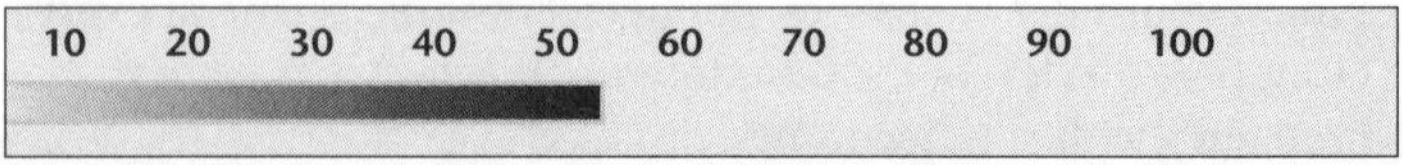

That's better than a dial. And nowadays it should show kilometres instead of miles, though I'm not much in favour of that.

C: Look, I'm not sure I can go along with all this. This speedometer looks like something off an Oldsmobile.

J: Never mind! Take a look at the horn. You got a separate pad there. Why not integrate. Just press the rim of the steering wheel. No effort! And the ignition key and switch. State of the art says you don't need a switch or button. You start the engine by turning the key. We got to make it a bit sporty. No round gear levers but a short racing lever etc

C: Hold on, this is a family car, you know.

J: Sure, but that doesn't mean you have to live in the last century. Listen, we've got to come up with some solutions. That comes direct from the Board of Management, from the CEO. So I suggest we do a brain-storming. Let's fix an appointment, say next Monday at 10 in the morning. We'll pool our ideas and then ... *(fade out)*

James' diary

Jesus, these Brits are over sensitive. Why did he get so uptight about the design? Anyone could see it was years out of date. Well, next week I'll be in the construction department. Wonder what'll be waiting me there.

T Working with the text

1. Are these statements true (T) or false (F)?

1. Today James is in the R & D Department. ☐
2. The director sent him there to help to get things off the ground. ☐
3. Mr Bennet thinks that American cars are uneconomical. ☐
4. Mr Bennet thinks that the new four-door range is fine. ☐
5. The new four-door range is a sports car. ☐
6. James wants to integrate the horn in the rim of the steering wheel. ☐
7. James thinks that speedometers with hands are old-fashioned. ☐

Food for thought

The marvels of modern technology include the development of a soda can which, when discarded, will last forever, and a $70,000 car which, when properly cared for, will rust away in two or three years.

Paul Harwitz

2. What they might have said but didn't

Replace the words in italics by those used in the text

1. He *thought* he knew all about the defects.	He .... he knew all about the defects.
2. The director sent me down to *gather* information.	The director sent me down to .... information.
3. Perhaps the American *method* can help to get things *started*.	Perhaps the American .... can help to get things ....
4. Your problem is *obvious*.	Your problem is .... in the face.
5. You don't *understand* me.	You're not .... my ....
6. *Wait* a minute!	.... a minute!
7. I don't *agree*.	I can't .... this.
8. We'll *solve this problem*.	We'll ....

H How things work in the car

Toolbox 1

speedometer
gear lever
dials
dashboard

1. Tools in use: Imagine you are a mechanic

Do you know where to put these parts? Match numbers and letters.

1 The speedometer is
2 The gear-lever is
3 The dials are
4 The instrument-panel is

A right of the driver's seat.
B in the instrument-panel.
C in the dashboard.
D in the speedometer.

Toolbox 2

petrol gauge	dashboard
bodywork	upholstery
instrument panel	dial

2. Tools in use: What is the 'it'?

You don't need a dictionary. The phrases below (situations) will help you.

It shows how much petrol you have left. ....
It is polished regularly in Germany. ....
It consists of several meters. ....
It has a glove compartment. ....
It makes the seat more comfortable. ....
It is a meter that shows your speed. ....
It is a meter that shows your mileage. ....

S Situations

1. At a petrol station: What are they talking about?

Match numbers and letters.

Match phrases and	parts of the car
1 Add some distilled water.	**A** battery
2 Could you wax it for me? It's a bit dull.	**B** tyres
3 The bulb's gone.	**C** paintwork
4 Could you give me some anti-freeze?	**D** headlight
5 They are worn. They need changing.	**E** radiator

2. In the Design Department: Let's refine the design

Put the adjective in the right place.

light, sporty, small, economical, modern

The old design	The new design	
is too big	must be	....
consumes too much petrol	must be more	....
is too heavy	must be	....
is a bit old-fashioned	must be more	....
looks like a family car	must look	....

Beef up your word power

1. What do you get when you put these together?

	nuts		plugs		water
....	tyre	....	piston	....	anti-freeze
	hubcap		valve		cells
	polish		jack		bonnet
....	wax	....	brace	....	boot
	wash		screwdriver		wing

2. Intelligent guessing

What's the German for the English words in italics?

1. The *bulb* in the headlight has gone. ....

2. The tyres are *worn*. They need changing. ....

3. There's a *dent* in the door. Can you fix it up? ....

4. Could you give me some *anti-freeze*? ....

5. Can you do a 10,000 mile *maintenance check*? ....

6. The *petrol gauge* is defective. ....

7. Do you have a *car wash*? ....

Our quiz for the experts

1. Who invented a combustion engine in 1880?

2. Which are the most famous British cars?

3. When did mass production of small and medium-sized cars begin in Europe?
A after the First World War
B in the nineteen thirties
C between 1900 and 1910

4. What did Ford invent in order to be able to mass produce cars.

5. Cars are less poisonous because they are fitted with
A exhaust silencers
B anti-exhaust killers
C catalytic converters

6. Name two of the big 'Motor Cities' in the USA.

7. Two devices make modern cars safer. Which?

8. Which fuel is not used for combustion engines?
A kerosene
B gasoline
C petroleum
D diesel

Time for a smile: Advice from the experts

1. When approaching a green light slow down and stop, in case René Bosewitz is coming the other way in his Saab.
2. If your brakes fail whilst driving at speed in the latest Ruver four-door model, release your bonnet catch. The raised bonnet will provide vital wind resistance and help slow down the vehicle.
3. If your brakes fail while reversing, open all your doors. The trees will bring your car to a standstill.

Ⓡ Extra reading: Car repair

Handy trouble-shooting chart

Problem	Cause	Solution
Suddenly the car stinks when engine is running.	A cat was sleeping on the engine.	No solution. Sell the car.
A horrible noise when car is moving.	a dog is tied to the rear bumper	Turn up the radio.
The car will not start.	Something is wrong with the car.	Tell your wife to push it.
The gearbox makes an awful noise when you change into third gear.	Something is wrong with the gearbox	Give the car to the poor and claim a tax reduction
The steering wheel comes off while driving.	Your son used it in a cross-country race.	Jump out. Give son a good beating.
The brakes fail while driving down a mountain.	You are pushing the wrong pedal, if not	Say a prayer.

Do you know your car?

The exhaust system: It is located under the car and smeared with road kills. From time to time you should drive briskly through a wading pool.
Tyres: Tyres are extremely important, for without them the tyre industry would cease to exist. You should inspect your tyres for the little letters and numbers on the side, which represent significant events in the lives of the tyre factory employees. For example, A78-13 means "All 78 of us tyre factory employees went out and got really drunk last night, so maybe 13 of the tyres we make today won't be any good."
The exterior: Hour after hour, day after day, month after month the sun beats down on your car with harmful rays that can fade the paint and damage your health if you spend any time outside trying to do anything about it. So the hell with the exterior!

Adapted from D. Barry, *The Taming of the Screw*. Rodale Press

C Let's do it crosswise

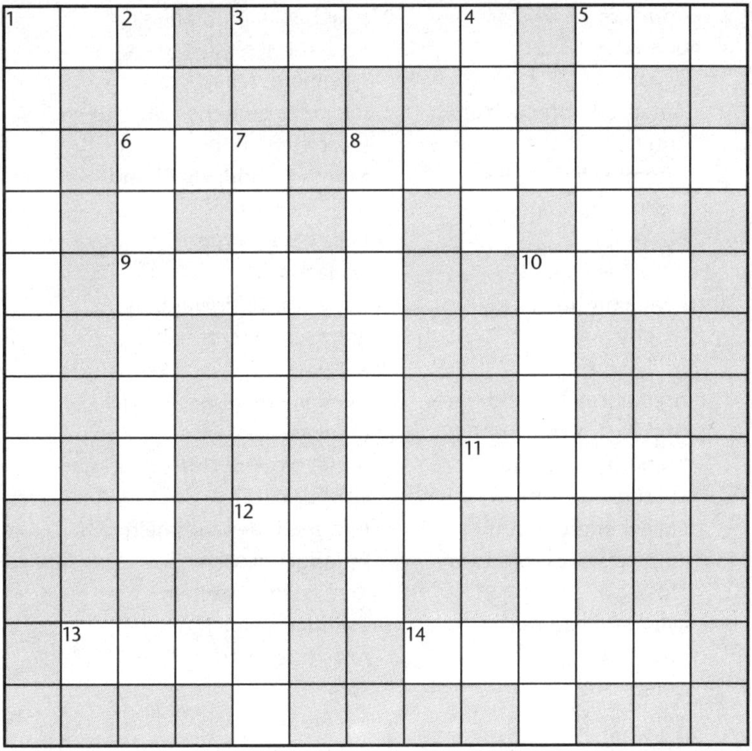

Across

1. English for 'Rand'
3. he went at a – ? – of 90 m.p.h.
5. American English for 'petrol'
6. opposite of 'high'
8. a hundred years
9. a gauge is a – ? –
10. you turn it to start the car.
11. he produced the Model T
12. sometimes the battery is – ? –
13. it should have a good profile
14. it contains several instruments

Down

1. it cools the engine
2. it shows the miles
4. English for 'Beule'
5. another word for 'meter'
7. put – ? – into the radiator
8. round
10. jet planes burn it
12. all combustion engines need it

Help: opposite = Gegenteil
contain = enthalten

Ⓥ Vocabulary

admit	zugeben, eingestehen
alterations	Änderungen
make minor ~	kleinere ~ machen
American approach	die amerikanische Methode
anti-freeze	Frostschutzmittel
assembly line	Montageband, Fließband
assemble components	Teile montieren
bar graph	Balkendiagramm
battlefield	Schlachtfeld
beat a problem	mit etwas fertig werden
bloody awful	ganz schön schrecklich
bodywork	Karosserie
bonnet catch; undo the ~	die Kühlerhaube entriegeln
brakes; put the ~ on	bremsen
bulb; the ~ has gone	Birne; die ~ ist durchgebrannt
century	Jahrhundert
collect information	Informationen sammeln
come up with solutions	Lösungen finden
cross; put a ~	Kreuz; ein ~ machen
cylinder head gasket	Zylinderkopfdichtung
dashboard	Armaturenbrett
defect	Fehler
defective	fehlerhaft
dent; put a ~ in the car	Beule; einbeulen
device	Gerät, Vorrichtung
dial	Uhr
distilled water	destilliertes Wasser
employ someone	jemanden ein-/anstellen
experience	erfahren, erleben
fail; the brakes ~	versagen; die Bremsen ~
famous	berühmt
fitted / equipped with	ausgerüstet mit
folks	umgangssprachl.: Leute, Kameraden
four-door range	viertürige Modellreihe

fuel	Kraftstoff, Brennstoff
gas	Benzin (US)
gas-guzzling	benzinschluckend
gasket, seal	Dichtung
gasoline (US)	Benzin
gauge	Meßinstrument
gear lever	Schaltknüppel
gents	umgangssprachlich: Leute
get something off the ground	etwas in Gang setzen
get uptight about	sich aufregen über
glove compartment	Handschuhfach
go along with	einverstanden sein
heavyweight tank	schwerer Panzer
highway (US), motorway (BE)	Autobahn
independent	unabhängig
indicator	Blinker
inspection; maintenance check	Inspektion
instrument panel	Instrumententafel
interior (n.)	Innenraum
interrupt	unterbrechen
invent	erfinden
maintain a car	ein Auto warten
maintenance	Wartung, Inspektion
medium-sized	mittelgroß
meter	Meßinstrument
milometer	Kilometerzähler
motorway	Autobahn (brit.)
oldsmobile	Oldtimer
overview; give an ~	Überblick; einen ~ geben
pad; brake-pad	Bremsklotz
paintwork	Lackierung
petrol gauge	Benzinuhr
poisonous	giftig
pool ideas	Ideen sammeln
provide	hier: ergeben
racing gear lever	sportlicher Schaltknüppel

radiator	Kühler
reckon	glauben, meinen, schätzen
refine a design	ein Design verfeinern
rev counter	Drehzahlmesser
reverse; put the car into the ~	rückwärts fahren; den Rückwärtsgang einlegen
rim	Rand
scrap	verschrotten
seal; gasket	Dichtung
separate (v.; adj.)	trennen; getrennt
space	Raum; hier: Land
speed	Geschwindigkeit
speedometer / speedo	Tachometer
sporty	sportlich
standstill; come to a ~	Stillstand; stehen bleiben
stare someone in the face	hier: jemanden in die Augen springen
styling department	Designabteilung
successful	erfolgreich
suggest	vorschlagen; anregen
suited to	geeignet für
swear; no need to ~	fluchen; kein Grund zu ~
switch	Schalter
switch the light on	das Licht einschalten
take something to heart	etwas übelnehmen
tank	Panzer
treat somebody as if ...	jemanden behandeln, als ob ...
turn the key	den Schlüssel umdrehen
tyre	Reifen
unbelievable	unglaublich
upholstery	Polsterung
view	hier: Ansicht
washer	Dichtscheibe
wind resistance	Luftwiderstand

Chapter 4

Inside the engine

- ✻ **OUR STORY: Engine trouble**
- T **Working with the text:** What they might have said but didn't
- H **How things work:** Inside the engine
- S **Situations:** Making suggestions at a meeting
  The complaint
- W **Beef up your word power:** Odd man out and intelligent guessing
- Q **Our quiz for the experts:**
- C **Let's do it crosswise:**
- V **Vocabulary:** Words about engines

Time for a smile

Why was the automobile mechanic fired?
He took too many brakes (and breaks).

Sticker on a Mini-Austin:
This was a Rolls-Royce before it went through the car wash.

⊛ Engine Trouble

James' diary

I'm sure making a name for myself around this company. Trouble is, it's as a trouble-maker rather than trouble-shooter. The problem is that the guys here are so fixed in their ways. They don't want to see any problems. Today it was the meeting to hammer out the matter of the gasoline and the diesel engines in the Land Ruver series. The marketing department collected thousands of complaints about worn parts, over-consumption of fuel, breakdowns, but these teams of researchers don't seem to want to hear it. They're like some sort of industrial civil servants. And, of course, I walked straight into the firing-line. Here's how it went:

James (**J**), *Dr. Novick* (**N**), *Mr Smith, senior engineer* (**S**)

J: *(fade in)*... and gentlemen, that's why I propose we look at our carburettor engine again and make some careful re-designs.

N: Mr Kelly, we made further developments on this engine in the last two years. There's not much left to do there.

J: I guess I must be wrongly informed then because I was told the Marketing Department had collected a list of complaints and claims as long as your arm.

N: Who put around such a ridiculous rumour?

J: Actually it was Mr Tooth, the Member of the Board.

N: Oh my god!

J: Yes, so we'd better get down to business! Now look at this engine! John, could you flash it up on the overhead? There gentlemen! Shall I start, or will you?

N: Mr Kelly, you can see clearly that there's not much wrong with that. I...

J: Excuse me, gentlemen, but look at it! Start at the top with accessories. The air filter! This type's not effective after 12,000 miles of use. It's old hat. And if the air's not clean, that'll cause other problems. And have you folks never heard of fuel injection? These old carburettors might be okay on vintage cars but not in 1998.

S: Okay, you've made your point. You don't need to be funny.

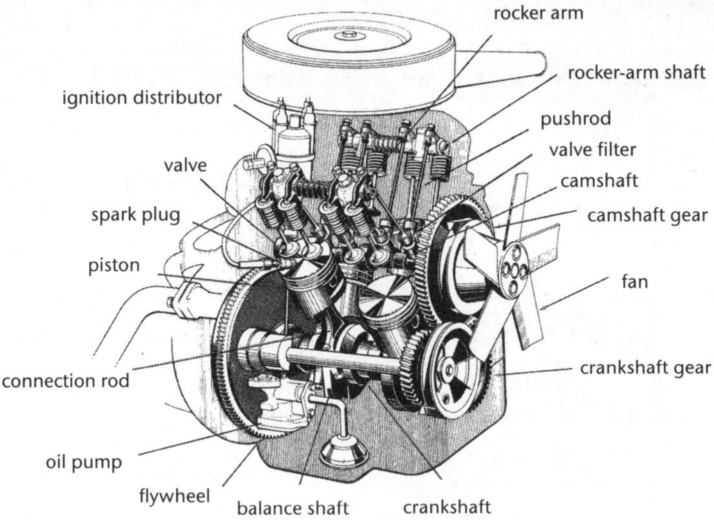

J: Okay, just my little joke. We've so many new developments to choose from. What about boxer motors for the sporty, 16 valves etc.? Why this old oversized motor? Wouldn't it be better to use new materials?

N: It's okay for the Land Ruver; it's an off-road vehicle.

J: Yes exactly, but off-road is sport. So we have to think of performance and looks ...

James' diary

I really felt I was fighting a losing battle. In the afternoon I had a brainstorming session with the Land-Ruver Diesel Department. I tried a different approach. They had to identify weak components or solutions themselves so the information didn't come from my side. Psychologically better. This was the result: they marked weaknesses with an arrow:
Well, I had to ask myself what we should actually keep. What on earth have they been doing in R&D in the last five years?

T Working with the text

What they might have said but didn't

Replace the words in italics by those used in the text

1. It was the meeting to *solve the problem* of the engine.	It was the meeting .... of the engine.
2. I *suggest* we look at our carburettor engine again.	I .... we look at our carburettor engine again.
3. *Let's* get down to business.	.... get down to business.
4. Could you *project* it on the OHP?	Could you .... it ...... on the OHP?
5. If the air's not clean, that'll *create* other problems.	If the air's not clean, that'll .... other problems.

H How things work inside the engine

Toolbox: Inside the engine

dip stick	dynamo
distributor	glower plug
carburettor	off-road car

Tools in use: What is 'it'?

You don't need a dictionary. Toolbox will help you.

It sends sparks to plugs. ....
It is long , oily and made of metal. ....
It is a small power station. ....
It transforms fuel into gas. ....
It is only in diesels. ....
It is a car for safaris. ....

Situations

1. Making suggestions at a meeting

Positive criticism

You shouldn't criticise people without offering a solution. One way of doing this is the suggestion. Let's look at the meeting again and see how James did it:

1. I *propose* (that)we look at our carburettor engine again and make some careful re-designs.
2. *We'd better* get down to business!
3. *Wouldn't it be better* to use new materials?
4. *What about* boxer motors for the sporty, 16 valves etc.?

Patchwork: More ways of making suggestions

Here are the patches for you to fill in:

wouldn't it		advice		advise
	submit		suggest	
recommend		might be		what about

1. May I .... a proposal?
2. My .... is to make the car look more sporty.
3. I strongly .... that we use a boxer motor.
4. .... changing the design?
5. Don't you think fuel injection .... a good idea?
6. .... be wiser not to use old materials?
7. I .... you to throw out all old components.
8. I would .... that we put off the decision.

Time for a smile: Stickers and buttons

A nuclear reactor is a safer place than Ted Kennedy's car.

Car window sticker in the USA

Don't ask me; I was hired for my looks

Button on a mechanic's T-shirt

2. Complaining to the Sales Department

Mick Disney (**M**), *Ruver Salesman* (**S**)

M: Hello, is that Ruver cars? I want to speak to somebody responsible.
S: Hello, this is Ruver Sales. How can I help you?
M: Well, for example you could start by keeping your part in the contract. I've never been so angry in my life. I want to complain most strongly.
S: What seems to be the problem?
M: Listen, it doesn't seem to be a problem. It is a problem. I bought a used Ruver Safari from you. I hadn't even got onto the motorway when the motor started backfiring. A few minutes later the exhaust pipe fell off.
S: Oh, I must apologise for that, sir. Are you sure you weren't driving off-road? Holes in the road and so on?
M: Look, there might have been a few roadworks but they didn't damage my exhaust. It's an off-road car after all, isn't it? And I'd like to complain about your after-sales service, or rather the lack of it. Your salesman promised me the car would do 45 miles to the gallon. It does 30. The engine's drinking too much. It's not running smooth. The ignition timing is out.
S: Teething problems. That can happen with a new series.
M: Just a minute! The car is two years old and already it is losing oil.
S: Oh, you bought a used car. A used car. Well, why didn't you say so? I'm only responsible for new cars. You want the used vehicle department. Ask them to overhaul the car from top to bottom. I'll put you through ...

What's wrong with Mick Disney's Ruver?

The first one has been done for you.
Here are a few words to help you with the diagnosis:

exhaust	distributor
carburettor	gasket

Causes	and their effects
Mick was driving off-road.	The exhaust pipe fell off.
The .... is out of time.	The motor doesn't run smooth.
There are holes in the	The motor backfires.
The .... needs setting.	The engine drinks too much.
The .... is worn.	The oil sump is leaking.

3. Technical defects require more than fixing

Learning by doing. Fit in the missing links from the words below:

accept angry apologise dissatisfied complaint
awfully sorry apologies complain

Complaint	and Apology
1 I've a .... to make.	I'm .... to hear that.
2 I really want to .... most strongly.	Please accept our ....
3 I'm utterly .... with the performance of my Ruver.	I can understand why you are ....
4 I really can't .... this kind of thing again.	Let me .... for the situation. I'll look into it immediately.

W Beef up your word power

1. Odd man out: What does not belong in the list?

A	B	C	D
breakdowns	replace	loose	anti-freeze
worn components	check	tight	exhaust fumes
over-consumption	change	worn	cooling liquid
broken parts	repair	noisy	gasoline

2. Intelligent guessing

Which parts of the diesel engine belong together? Match them.

1	piston	A	glower plugs
2	radiator	B	ring
3	dynamo	C	atomiser
4	compression chamber	D	cooling liquid
5	carburettor	E	cable

Q Our quiz for the experts

1. Which device avoids pollution of the air?
   A catalytic converter
   B silencer
   C float chamber
2. True or False? In a Wankel engine the pistons rotate instead of moving up and down.
3. What is the shape of a Wankel piston?
   A round B triangular C oval
4. The car's worst enemy is
   A lubrication B friction C rust
5. What is the effect of oil in an engine?
   A lubrication B purification C conservation
6. If you press the accelerator you open the
   A valve B nozzle C throttle
7. What is injected into the cylinder?
   A droplets of petrol B vapour of air and fuel
   C a mixture of air and fumes

Time for a smile

Ruver made in the UK:
absolutely safe and reliable; approved to British Standards BS 9966
(but so was the Titanic)

C Let's do it crosswise

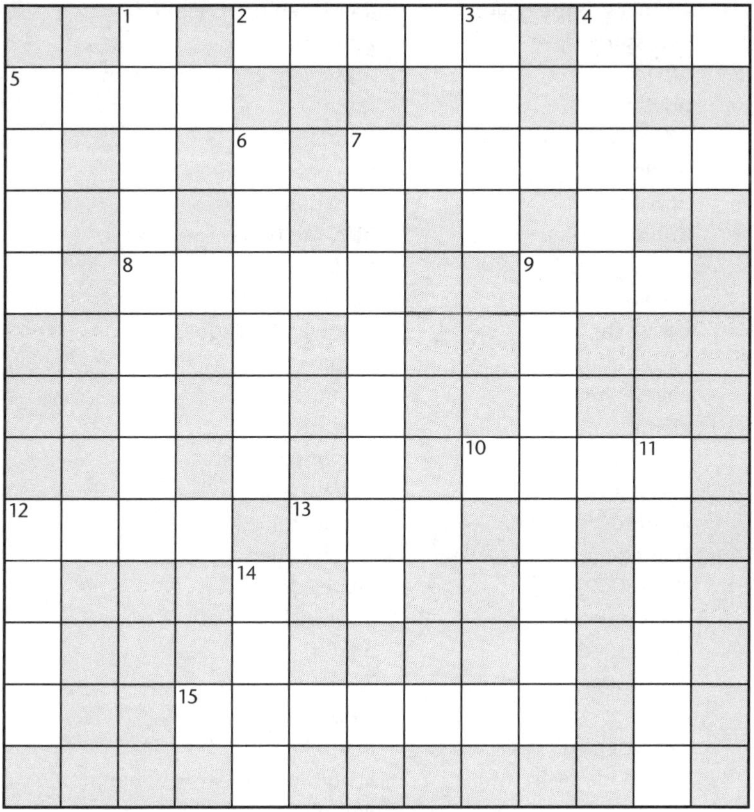

Across

2. miles or kilometres per hour
4. American English for petrol
5. opposite of 'succeed'
7. a hundred years
8. any instrument in the panel
9. you turn it to start the car
10. he first used assembly lines
12. it's black and made of rubber
13. device to cool the radiator
15. go backwards

Down

1. it shows you the distance
3. English for 'Beule'
4. another word for 'meter'
5. combustion engines burn it
6. British English for 'gas'
7. round
9. fuel for jets
11. English for 'Gerät, Apparat'
12. you do it with a key
14. car manufacturer in Germany

Ⓥ Vocabulary

45 miles to the gallon	45 Meilen pro Gallone
accessories	Zubehör
advice; give ~	Ratschlag; einen Rat geben
advise	raten
after sales service	Kundendienst
apologise	sich entschuldigen
apology	Entschuldigung
approach (n.)	hier: Methode
arrow	Pfeil
break	Pause
car wash	Autowaschanlage
carburettor	Vergaser
civil servant	Beamter
claim	Anspruch
complain	sich beschweren
complaint	Beschwerde
compression chamber	Verdichtungskammer
damage (v.)	beschädigen
dip stick	Ölmeßstab
disappointed	enttäuscht
dissatisfied	unzufrieden
distributor	Zündverteiler
droplets	Tröpfchen
exhaust pipe	Auspuffrohr
firing-line; be in the ~	Schußlinie; in der ~ stehen
friction	Reibung
fuel	Treibstoff
fuel injection	Treibstoffeinspritzung
fumes	Abgase
further development	Weiterentwicklung
gasket	Dichtung
gents (gentlemen)	meine Herren
glower plug	Glühkerze

guy	Bursche, Kerl
hammer out	ausbügeln
hole	Loch
keep a contract	einen Vertrag einhalten
looks	Aussehen, Äußeres
lubrication	Schmierung
make one's point	seine Meinung sagen
Member of the Board	Vorstandsmitglied
nozzle	Düse
off-road; drive ~	im Gelände fahren
out of time	unrund, falsch getaktet
over-consumption	überhöhter Benzinverbrauch
performance	Leistung
recommend	empfehlen
re-design (v.); make a ~	überarbeiten
replace	ersetzen
researcher	Forscher
responsible	verantwortlich
ridiculous	lächerlich
rumour; a ~ goes round	Gerücht; ein ~ geht um
running smooth	rund, ruhig laufen
session	Sitzung
soft-top / convertible	Kabriolett
submit a proposal	einen Vorschlag unterbreiten
teething problems	Kinderkrankheiten
throttle	Drosselklappe
valve	Ventil
vapour	Dampf
vintage car	Oldtimer
weakness	Schwäche, Schwachpunkt
worn parts	abgenutzte Teile
wouldn't it be wiser to ...?	wäre es nicht klüger zu ...?

5 Chapter

Looking at production tools

Time for a smile: Murphy's laws of the workshop

1. It works better when you plug it in.
2. When all else fails, read the instructions.
3. The most delicate object will be the one to fall on the floor.
4. Any object will fall so as to do the most damage.
5. Experience grows directly with equipment ruined.

(*) Bored in the workshop

James' diary

I seem to have gotten myself into hot water in the last weeks. Anyway a lot of processes and methods need to be changed around here if quality targets are to be reached. And that is also true for the shop-floor and assembly lines and the manufacturing in general. I got a feeling for how bad the situation is last Monday. Our team of trainees was called together to be shown around the workshops. The theme was 'manufacturing processes'. Our guide was the assistant to the Board.

Dr. Riser (**R**), *James Kelly* (**J**), *three trainees* (**T 1–3**)

R: And now, Gentlemen, we come to shop floor. We also use it for training our apprentices. I'd like to give you an overview of what we're using at present. If you have any questions, don't hesitate to interrupt me. Please follow me. Over here you see our high-speed lathe:

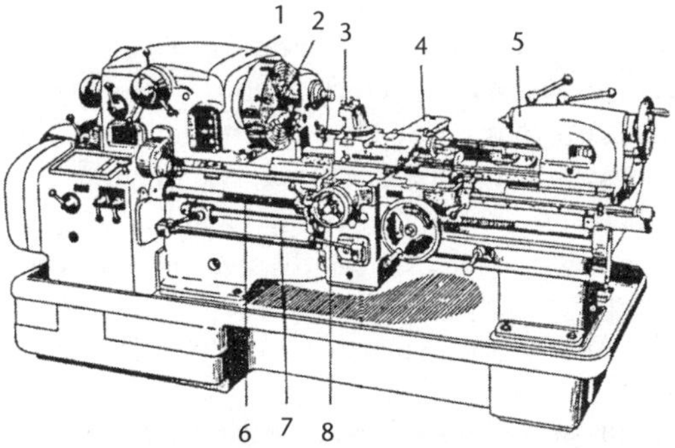

1 headstock
2 face plate
3 tool post
4 cross slide
5 tailstock
6 lead screw
7 feed shaft
8 feed control

As you can see, there are some important parts here including the gear box, tool-holder, the lever for normal and coarse threads and some other things. And here are the turning or cutting tools we use.

T1: Excuse me for interrupting you, Dr. Reiser, what exactly are these threads used for?

R: My goodness! Don't you know? They are for cutting into metal parts. and now ...

T2: Sorry for breaking in again, Dr. Reiser, but may I ask a question, too? You mentioned the turning tools. What are they needed for?

R: Well, young man, I'd have explained it if I'd had more time. The piece to be worked is fixed to this spindle. It is turned by the lathe. Next we move a machine tool towards it. In addition we have several other types of metal-working machines like this milling machine, for example. We are proud to possess such sophisticated equipment ... *(fade-out)*

J: *(to himself)* Jesus Christ. Where did they dig up all this ancient equipment? No automation at all! And where did they find this professorial Dr. Climber or whatever his name is?

R: *(continuing)* ... these include a grinding machine, a shaping machine, metal cutting circular saw, a drilling plant ...

T3: Pardon me for interrupting you again. A shaping machine? How interesting. Which parts are shaped on this tool?

R: Any parts, my dear fellow, that require a special form.

T1: Sir, sorry to break in... Could you describe some of the logistics machinery in our shops?

R: Since you ask, I'll include that information now instead of later as I had planned. We do indeed have an erection shop ... No, my young friend, it's not what's on your mind. Please follow me. Here they build up the various types of tooling machines. The shop has several workshop cranes such as overhead cranes and fixed hoisting or lifting gear such as this bracket crane that is fitted to the wall. The jib or crane arm protrudes ...

J: *(interrupting)* Excuse me, but may I just draw your attention to those gas bottles over there.

R: Yes, my foreign friend, those are for welding with the oxygen-acetylene gas torch. It won't work without the gas even in America.

J: Yes sir, you're right but we don't use them in an enclosed space. Danger of explosion. What does the safety committee say about this?

R: I'm sure I wouldn't know. Now, if we can get back to the more important matters. Please look once again at ... *(fade-out)*.

James' diary

What a lot of technical hogwash! Wasting our time making a tour of an out-of-date plant, watching old welding, soldering, casting and moulding processes. It's time they looked at good suppliers who can deliver the basic engine and other parts cheaper and in better quality. And that Dr. Riser thinks he's still lecturing at university. In the end he forbade me to ask questions. It must be the island mentality. To be fair, though, there was one interesting part to the whole presentation. That was the tour around the bodywork assembly and welding area. They've got a robot line. I must see Dr. Block about it because in our company they are only used for painting and joining parts of bodies together. In some other companies they do almost all the work on the assembly line. Our worker-robots need to be given new instructions so they can do more intelligent work. Come to think of it, perhaps robots could replace some of the engineers here. They might do the job better.

T Working with the text

1. Are these statements true (T) or false (F)

1. Ruver produces most of the components themselves. ☐
2. Tools are adapted to new designs in the erection shop. ☐
3. Mud-guards are formed on shaping machines. ☐
4. Coarse threads are manufactured on the lathe. ☐
5. Welders use gas torches. ☐

Time for a smile: The law of the workshop

Any tool, when dropped, will roll into the least accessible corner of the workshop.

If it jams – force it. If it breaks, it needed replacing anyway.

2. What they might have said but didn't

Replace the words in italics by those used in the text

1. I'd like to give you an *idea* of what we're using.	I'd like to give you an .... of what we're using.
2. Where did they *find* all this ancient equipment?	Where did they .... all this ancient equipment?
3. *Excuse me, but can you see* those gas bottles over there?	May I .... to those gas bottles over there?
4. We are proud to *have* such sophisticated equipment.	We are proud to .... such sophisticated equipment.
5. It won't *function* without the gas.	It won't .... without the gas.
6. If we can *return* to the more important matters.	If we can .... to the more important matters.

H How things work

Toolbox 1

1 lathe **2** shaping machine **3** mould
4 bracket crane **5** drilling machine

1. Tools in use 1: Match numbers and letters

1 drilling machine	**A** It shapes a workpiece by turning it.
2 mould	**B** It gives sheet iron its final form.
3 shaping machine	**C** You pour liquid metal in it.
4 bracket crane	**D** You make small holes with it.
5 lathe	**E** You lift heavy objects with it.

Time for a smile: The Law of the workshop

1. When you try to prove that your machine doesn't work, it will.
2. The work-bench is always untidier than last time.

2. Shops that don't sell anything

A shop floor **B** erection shop **C** test shop **D** workshop

In which shop would you put the following posters?
The first has been done for you.

1. The one time in the day you lean back and relax is the one time the Boss walks in?	A
2. If it were be cheaper to buy a new unit, the company would insist on repairing the old one.	
3. The motor will rotate in the wrong direction.	
4. When an error has been detected and corrected, it will be found to have been correct in the first place.	
5. After the machine tool has been assembled, extra components will be found on the bench.	
6. If a machine tool finally functions perfectly all the other production units will malfunction.	
7. Any wheelnut, when dropped will roll into the least accessible corner.	
8. After the last of 16 screws has been tightened, it will be discovered that the gasket has been forgotten.	

Toolbox 2

blowtorch	Lötlampe
soldering iron	Lötkolben
milling machine	Fräse
welding rod	Schweißelektrode

3. Tools in use 2: What is the 'it'?

You don't need a dictionary. The following phrases will help you.

It is used to give a sheet of iron a form. ....

It is a machine to make pistons with. ....

It joins parts with an alloy of tin and lead. ....

It uses electricity to join larger parts. ....

Ⓢ Situations: How to interrupt

Dr. Riser encouraged his audience to interrupt him if they had a question. Here are some useful phrases for interrupting and continuing with one's talk after being interrupted. Learn them by "assembling" them:

break in		finish		interrupting
	return to		continue	
interrupt		coming		go back

May I .... you for a moment?
Pardon me for............................, but is this really relevant?
Excuse me, could I just .... here?
If I could just ...., I'll answer your question in a minute.
.... back to what I was trying to explain ...
To .... to what I was saying ...
To .... my subject ...
If I might just .... my sentence.

Ⓦ Beef up your word power

1. What do you get when you assemble the components?

If you don't know, have a look at our quiz for the experts

	handle		grip		handle
....	head	......	blade	....	blade
	claw		teeth		sharp edge

2. The Ostfriesen final exam for engineers

1	What do you dig holes with in the garden?	**A**	hammer
2	What do you chop wood with?	**B**	level
3	What do you measure your height with?	**C**	rule
4	What do you drive in nails with?	**D**	axe
5	What do you need to check surface level?	**E**	spade

Our quiz for the experts

The untidy workbench

The following tools are lying around on expert's workbench. Name the tools.

spanner hammer drill gloves trowel jig saw
pliers screw driver

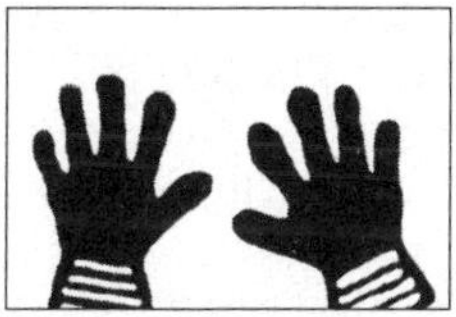

1 ....

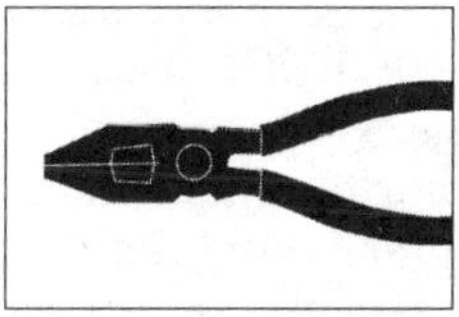

2 ....

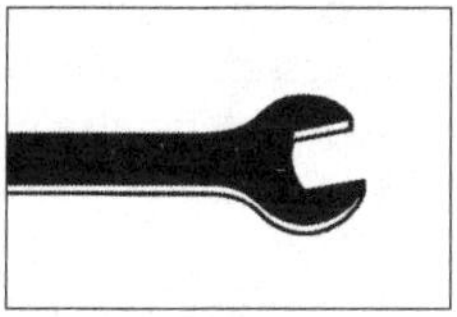

3 ....

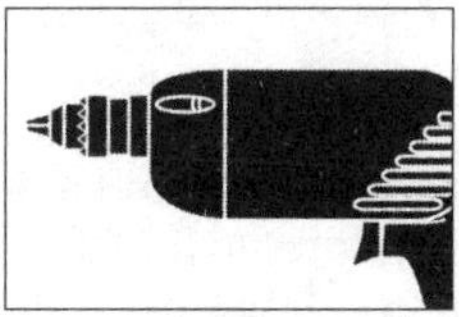

4 ....

5 ....

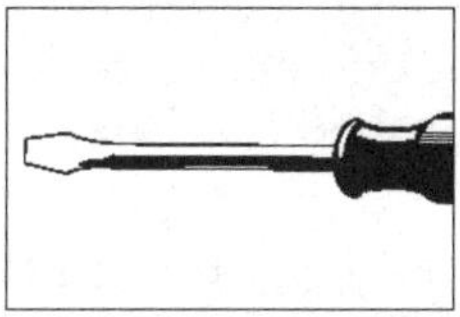

6 ....

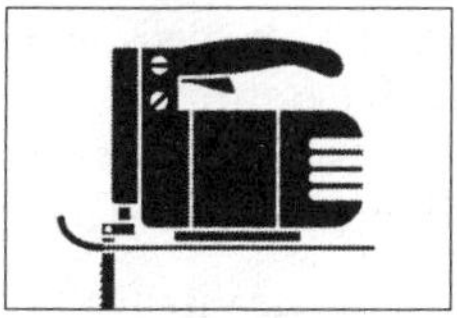

7 ....

8 ....

Ⓡ Extra reading: All you should know about plumbing

You should always be worrying about your plumbing. Take the tragic story of my friend, René and his girlfriend who went on a holiday to Barbados not knowing that one of their pipes had sprung a leak. By the time they returned, water had destroyed home and office forcing them to collect 250,000 marks from the insurance company and use the money to go back to Barbados for an unlimited holiday. All they are doing now is lying in the sun sipping tropical rum drinks.
This needless tragedy would never have occurred if the couple had taken more interest in their plumbing which requires only a few simple tools and the willingness to stick your arm into a clogged toilet after a diseased houseguest has used it. Well, René is not a hobby plumber. When an annoying faucet is dripping, all he does is turn up the radio, although basic plumbing techniques do not require much intelligence.
Let's look at how plumbing works.
A plumbing system is very much like your electrical system, except instead of electricity, it has water, and instead of wires, it has pipes, and instead of radios and toasters, it has taps and toilets. So the truth is that your plumbing system is nothing at all like your electrical system, which is good, because electricity can kill you. The major problem with plumbing systems is that they tend to leak. It usually happens on a Sunday afternoon. Don't try to understand why.
The key to preventing leaks is maintenance. At least once a year you should go around and hit the various elements of your plumbing system with a hammer. If water comes pouring out, don't be afraid to tackle it yourself. The only difference between you and an experienced master plumber is that he will charge you a huge bill for it and you can't charge anybody.

1. Phone your architect and ask him where the valve is that turns the water off. It is usually covered with slime and spiderweb.
2. Take a walking stick and kill the spider.
3. Use a wrench when you can't turn the valve handle because it is completely rusted. Turn it clockwise till it breaks off.
4. Replace the valve and the leaking pipe.
5. Phone the experienced master plumber to turn the water back on .

By the way, phoning the experienced master plumber mightn't be such a good idea. As a rule, leaks occur on Sunday afternoons.

C Let's do it crosswise

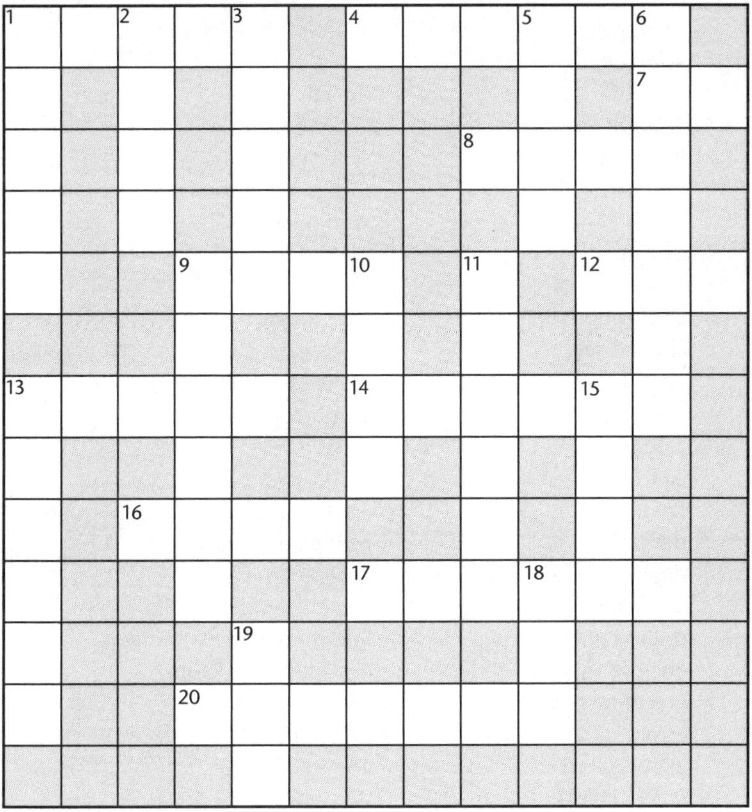

Across

1 machine tool
4 it's annoying when dripping (USA)
7 the opposite of 'off'
8 put together (parts, components)
9 handle of a saw
12 you do it with a knife
13 you do it with a crane
14 English for 'Flachzange'
16 pipes tend to do it sometimes
17 he uses heat to join metal parts
20 pouring liquid metal into moulds

Down

1 it tells you if it's horizontal
2 the opposite of 'untidy'
3 mistake
5 you do it with an axe
6 dentists have it too
9 it keeps the oil in the oil pan
10 part of your plumbing system
11 you hit it with a hammer
13 part of the hammer
15 it tells the length and width
18 you do it with a spade
19 its blade has got teeth on it

Ⓥ Vocabulary

accessible	zugänglich, erreichbar
addition; in ~	zusätzlich
ancient	(ur-)alt
as a rule	in der Regel
blowtorch	Schweißlampe
bodywork assembly	Fahrgestellmontage
bracket crane	Wandkran
casting	Gießen
chisel	Meißel
chop wood	Holz hacken
circular saw	Kreissäge
claw	Kralle (des Hammers)
clockwise	im Uhrzeigersinn
clogged	verstopft
coarse	grob
crane	Kran
delicate	zerbrechlich, zart
dig up	ausgraben
diseased	krank
drilling plant	Bohrmaschine, Bohranlage
enclosed space	geschlossener Raum
erection shop	Werkzeugmontage
error	Fehler, Irrtum
experienced	erfahren
faucet (USA)	Wasserhahn
gas torch	Lötlampe
gasket	Dichtung (Ölwannen~)
get into hot water	in Schwierigkeiten geraten
grinding machine	Schleifmaschine
grip	Griff
handle	Stiel, Griff
head	Kopf (des Hammers)
hogwash	Quatsch

hoisting or lifting gear	Hebevorrichtung
including	einschließlich
insist on	bestehen auf
instead of	anstatt
jib	Ausleger (eines Krans)
join parts together	Teile zusammenfügen
lathe	Drehmaschine; Drehbank
least accessible	am schwersten zugänglich
lecture	Vortrag, Vorlesung
level	Wasserwaage
logistics	Logistik
maintenance	Wartung
major problem	Hauptproblem
malfunction	nicht / schlecht funktionieren
manufacturing	Herstellung, Produktion
moulding	Formen
mud guard	Kotflügel
needless	unnötig, überflüssig
pipe	Rohr
plane	Hobel
pliers	Flachzange
plug in	einstecken (den Stecker)
plumbing	Klempnerei
prevent	verhindern, vorbeugen
professorial	professoral, gelehrt
protrude	herausragen
quality targets	Qualitätsziele
robot line	Roboterstraße
rule	Meterstab
safety committee	Sicherheitsrat
shaping machine	Formpresse
sheet iron	Blech
shop-floor	Werkstatt (allgemein)
slime	Schleim
soldering	Löten

soldering iron	Lötkolben
sophisticated equipment	hochentwickelte Ausrüstung
spiderweb	Spinnwebe
spring a leak	ein Leck bekommen
tackle	in Angriff nehmen
tap	Wasserhahn
thread	Gewinde
tongues	Beißzange
torch	(Löt-)Lampe
turning	Drehen
untidy	unordentlich
welder	Schweißer
welding process	Schweißvorgang
wire	Draht
work-bench	Werkbank
workpiece	Werkstück
workshop	Werkstatt, Montage
wrench	Schraubenschlüssel

Electricity and all that hot stuff

- (*) **OUR STORY: A Sumo-style attack on Ruver**
- (T) **Working with the text:** Did you get the message?
- (H) **How things work:** The power-plant
- (S) **Situations:** Phrases for presentations
- (W) **Beef up your word power:** Power-plant energy
- (R) **Extra Reading:** All you ever wanted to know about electricity
- (Q) **Our quiz for the experts:** A short history of electricity
- (V) **Vocabulary:** Words about energy

Time for a smile: These incredible Japanese

Sumo wrestling is the survival of the fattest.

The Italians make the best lovers,
but the Japanese make them smaller and cheaper.

⊛ A Sumo-style attack

James' diary

I've had the pleasure of working for Ruver for more than six months now. Well, I've got some things moving all right, but I'm none too popular here. Anyway I was getting used to things, starting to take it easy, when 'bang' out of the blue: restructuring. The Japanese multi Gonda is said to have bought up 35% of shares or even more. They seem to be involved at all the higher levels and especially in manufacturing management. Here are the figures Dr. Reiser gave me the other day:

Percentage of Japanese participation in committees			
(50%)	(50%)	(30%)	(30%)
Production Management	Works Engineering	Quality Control	Labour Relations

I don't know how they managed it, but the Japanese must have landed a good contract in this joint-manufacturing enterprise. And the next big shock was the following:

Jean Wotnot (James' secretary) (**W**), *James Kelly* (**J**)

W: Mr Kelly, you're to attend a meeting at 10 a.m. sharp this morning.
J: Hey, that's not planned in.
W: It's Mr Okanawi, director of engines.
J: Well, he's certainly a big fish. All Ruver engines will be replaced by ones from Gonda. They're better quality in any case. Okay, I'll be there.

James' diary

Well, bless my booties! I had a half discussion or meeting with Mr Okanawi where he asked me very politely to describe, project by project, what I'd been doing since my arrival at Ruver. You know what? He seemed to be asleep. I coughed a few times to wake him but nothing happened. After a while he opened his eyes, thanked me and informed me they had chosen me to go to Indonesia to check on some problem with their power plant. It seems the Ruver subsidiary in Indonesia produces various metal components, exhausts, housings and even gears. The problem is that they suddenly had no power. Result: shortages of parts in Ruver, England. I was on the next plane.

Meeting at Ruver subsidiary, Indonesia. James Kelly (**J**), *Mr Maylay, Senior Production Manager* (**M**)

J: Well, I hope you'll be able to describe the problem in a few simple sentences for a layman like myself.

M: I have full confidence in your superior European know-how, Mr Kelly.

J: Well, thank-you for your support, but I'm American actually, from Ohio. Could you explain the matter to me?

M: It's very simple really. You know there are several types of nuclear power plants all using nuclear fission. Ours was one of the first boiling-water reactors in this part of the world. If you'll allow me, I'll guide you through the stages. Don't hesitate to interrupt me when you have a question. *(Mr Maylay explains a flow-chart in a very precise manner).*

M: That's how the power is produced in Indonesia, which has given us the energy to melt our ores from which we cast the many parts which your headquarters buys far cheaper than they could in Europe. That's capitalism!

J: Actually, Mr Maylay, that's business, as we understand it. So what exactly is the problem?

M: The problem is that our government is going to close the nuclear plant by June, 1999. They have obviously had serious problems with it since the last earthquake, too much radioactive pollution of the air and the river.

J: Well, why did you explain to me how your reactor works if they're going to scrap it.

M: There is a rumour that the government is going to invest in solar, wind and tidal plants rather than modernising the reactor. We must search for other sources of energy at the appropriate price. We need a hydroelectric power station near the factory that uses those huge waterfalls you can see from the window. That's why you are here.

J: Actually, I'm an car expert. I ...

M: *(interrupting)* You are the representative of your company, your automobile company. Gonda has great influence with important decision-makers in this country. After all, they employ more than 25,000 workers here.We look forward to your suggestions ...

Working with the text

1. Are these statements true (T) or false (F)?

1. Ruver and Gonda have negotiated a joint venture. ☐
2. Ruver engines will be replaced by ones from Gonda. ☐
3. James is sent to Indonesia to buy parts for Gonda engines. ☐
4. The Indonesian reactor uses nuclear fusion. ☐
5. Mr Maylay expects James to talk to the government. ☐
6. He thinks of building their own nuclear power plant. ☐

2. What they might have said but didn't

Replace the words in italics by those used in the text

1. Mr Kelly, you *have* to attend a meeting at 10.	Mr Kelly, you .... to attend a meeting at 10 a.m.
2. All engines *are to* be replaced by ones from Gonda.	All engines .... be replaced by ones from Gonda.
3. Thank-you for your *help.*	Thank-you for your ....
4. That's *the way* the power is produced in Indonesia.	That's .... the power is produced in Indonesia.
5. We *have to look* for other sources of energy.	We .... for other sources of energy.
6. That's *the reason for your presence.*	That's ....

H How things work in the power plant

God has created the wind, the waves, the waterfalls, the rivers, the magnetic North and South Poles. That is all you need to produce electricity.

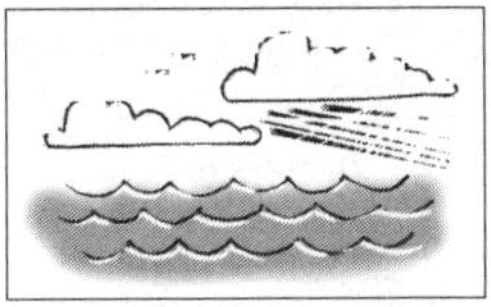

An ancient Greek invented the water wheel, another the wind wheel. Get the wheel into a stream of water, steam or wind and it will turn around.

Fix a coil made of wire to the end of the wheel's axle. Let it spin between the poles of a U-shaped magnet.

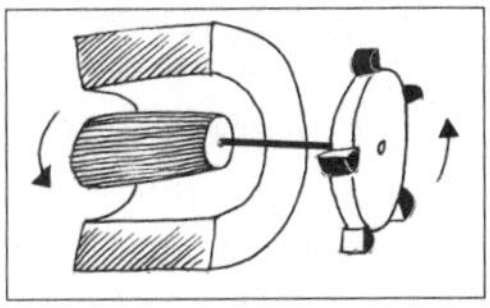

The electrons in the wire don't like it when the wire cuts through the magnet's lines of force. They get nervous and begin to move. There you have your electric current.

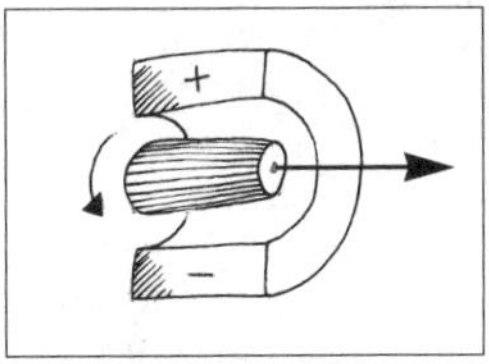

Tool box 1

In a modern power plant things have more complicated names:

chute	turbine
generator	coil

1. Can you match these parts with the things in our story?

1 wheel	A coil
2 wire ring	B chute
3 coil and magnet	C turbine
4 waterfall	D generator

Tool box 2

power plant	voltage	transformer(s)
power line(s)	alternating current	coil(s)
pylon(s)	magnetic field	volt (s)
supply	distribution	turn(s)

2. How the electricity travels to your home

Fit in the words from Tool box 2

The electric current travels from the **(1)**..................... to your home through a **(2)** .... suspended on high **(3)** .... On both ends there are **(4)** .... Why? Well, the current should travel at high **(5)** .... through the **(6)** .... At a low voltage a lot of energy would be lost on the way. That's where the **(7)** .... comes in.
A **(8)**.................... changes the voltage of an **(9)** .... It is an iron core with **(10)** .... wound around on both sides. The alternating input current switches a **(11)** .... on and off which is transferred to the output **(12)** .... The input coil has fewer **(13)** ....than the output coil. The change in voltage depends on the difference of **(14)** .... in the two coils.

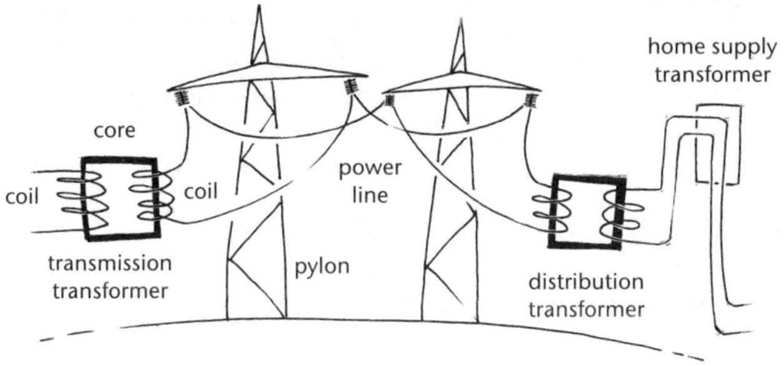

The transmission transformer steps the (**15**) .... up to several hundred thousand (**16**) .... The (**17**) .... transformer steps it down to several thousand volts and the home (**18**) .... transformer steps it down to 220 volts. And that's what you get in your home!

Situations: Phrases for presentations

Phrases Mr Maylay might have used

Put these patches back into the right place.

overview		interrupt		appreciate
	involved		topic	
main parts		summarise		attention
	upper part		glad	

1. I'm .... to have the chance to speak to you about ...
2. I .... that you're sacrificing your valuable time to listen to me.
3. Don't hesitate to .... me when you have a question.
4. The .... of this presentation is ...
5. Let me first give you an .... of the situation.
6. My presentation will be in three ....
7. I'm going to explain the problems .... in getting electricity.

8. May I draw your .... to the handout in front of you?
9. So to .... we can say that ...
10. The .... of the slide shows the interior of the container.

Beef up your word power

1. What do you get when you assemble the components?

....	core voltage coil
....	iron core coil magnet
....	chute turbine generator
....	control rods uranium pellets coolant pump

2. Words around a nuclear power plant

Match definitions	and parts of the PWR
1 It contains rods of uranium.	A water circuits
2 They stop the chain-reaction.	B reactor core
3 It should survive most accidents intact.	C control rods
4 It prevents water from turning to steam.	D containment
5 They prevent the overheating of the core.	E pipes
6 They take high pressure stream to the turbine.	F pressuriser

Time for a smile

Ronny told Annette that there was a chance that radiation might hurt his reproductive organs but she said in her opinion it was a small price to pay.

"What's all that fuss about plutonium?" asked Annette. "How can something named after a Disney character be dangerous?"

Our quiz for the experts

A short history of technology

Are these incredible statements true or false?

1. Large-scale steel production started in the USA in 1850 after the steel converter was invented by (James Kelly's grandfather) William Kelly. ☐
2. Gotlieb Daimler fitted his first petrol engine to a boat in 1873. ☐
3. He fitted his second motor to a bicycle in 1875. ☐
4. Is it true that a Greek by the name of Hero invented the first steam engine? ☐
5. Thomas Alva Edison invented the electric light? ☐
6. The rotor blades of wind turbines are huge – up to 150 meters. ☐
7. The nuclear reaction that happens inside a reactor is called nuclear fusion. ☐
8. A nuclear reactor converts one form of energy into another. ☐
9. PWR is short for pressurised water reactor. ☐
10. The control rods in a reactor are made of plutonium. ☐

Time for a smile

British manufacturers can't win. They produce a new machine and within three months the Russians have invented it, and a month later the Japanese are making it for half the price.

Cut electricity bills in half by changing the flex on all your electrical appliances for a much thinner one which will allow only half of the electricity through.

Ⓡ Extra reading: All you ever wanted to know about electricity

Have you ever wondered what the electrical system at home is all about? Well, it is very much like your plumbing system, except instead of water, it has electricity, instead of pipes, it has wires, and instead of taps, it has switches and fuses. So the truth is that your electrical system is nothing at all like your plumbing system, which is good, because broken pipes destroy your home.

Electricity is made up of extremely tiny creatures, known as electrons which you cannot see with the naked eye unless you've been drinking a lot. They travel at the speed of light, which in most European homes is 220 volts per hour. That is extremely fast. In the time it has taken you to read this sentence, an electron, could have travelled all the way from Hollerberg to Gelsenkirchen and back , if it decided to do so.

When electrons travel in company they do it neither in pipes nor on roads but in wires. When you join all the electric wires in your home together correctly you build a circuit which brings the electrical current into your home and out again, before it has a chance to kill you. If you don't do it properly you get a short circuit. Electrons don't like short circuits. Lengthen it or they will burst out in form of sparks, damage your carpet or go on strike which is known as a blackout. That's where fuses come in. They keep the electrons under control as long as you change them regularly.

The five best known currents are alternating current, direct current, lightning and the horrible Anglo-Saxon-flickering current. Most European homes have alternating current, which means that the electron goes in one direction for a while, changes its mind and goes in the other direction. That prevents a harmful traffic jam in the wires. The Anglo-Saxon-flickering current is what you get in Scottish castles, which as you know, are often haunted by ghosts. You plug in your electric razor, the light flickers and all the escalators in town stop.

When the light flickers in your home you might have some ghosts, too. Don't worry, read all about it in the fine documentary on The Black Exorcist of Dartmoore Dungeon or call in an electrician licensed by the Vatican to do the trick.

(V) Vocabulary

alternating current(AC)	Wechselstrom
appreciate	zu schätzen wissen
appropriate price	angemessener Preis
attend a meeting	an einer Besprechung teilnehmen
attention	Aufmerksamkeit
bless my booties	Oh, mein Gott
blow me down!	das haut mich um!
boiling-water reactor	Siedewasserreaktor
cast; ~ parts for engines	gießen; Motorteile gießen
chain-reaction	Kettenreaktion
chute	hier: Fallrohr; Rutsche
circuit	(Strom-)Kreis
coil	Spule, Schleife
confidence	Vertrauen
containment	Behälter, Mantel, Schale
cough	husten
direct current (DC)	Gleichstrom
distribution	Verteilung
earthquake	Erdbeben
electric current	elektrischer Strom
flex	Stromkabel
flicker	flackern
fuse	Sicherung
fuss	Aufregung
get used to	sich gewöhnen an
Greek	Grieche, griechisch
haunted by ghosts	von Geistern heimgesucht, verwunschen
hesitate	zögern
housing	Gehäuse
iron core	Eisenkern
joint venture	Gemeinschaftsunternehmen
labour relations	Arbeitgeber-Arbeitnehmer-beziehungen
layman	Laie

light bulb	Glühbirne
lines of force	Kraftlinien
nuclear fission	Atomspaltung
nuclear fusion	Atomkernverschmelzung
ore	Erz
overview	Überblick
participation	Teilnahme, Beteiligung
pollution	Verschmutzung
popular	beliebt
power line	Hochspannungsleitung
power plant	Elektrizitätswerk
power supply	Energieversorgung
prevent	verhindern
radiation	Strahlung
reproductive organ	Zeugungsorgan
restructuring	Umstrukturierung
rod	Stab
rotor blade	Rotorblatt
scrap	verschrotten
short circuit	Kurzschluß
shortage	Engpaß, Knappheit
slide	Dia; Folie
sources of energy	Energiequellen
step up / down	hoch- /herunterspannen
stocks /share	Anteile / Aktie,
subsidiary	Tochtergesellschaft
summarise	zusammenfassen
superior (adj.)	überlegen
survival	Überleben
suspend	aufhängen
switch	Schalter
tap	Wasserhahn
thank-you for your support	hier: vielen Dank für Ihre ermutigenden Worte
transformer	Transformator
uranium pellets	Uranscheiben, -würfel
vanish	verschwinden
wrestling	Ringen

How things work in phones

Time for a smile: Communication problems

The phone rang in the maternity ward:
Nurse: St Joseph's hospital. Nurse Cassandra speaking.
Man: I'm bringing my wife – she' s going to have a baby. It's urgent.
Nurse: Is this her first baby?
Man: No, no. This is her husband!

⊛ The art of communicating

James' diary

I came to Europe as a glorified car mechanic, well an automobile expert, and it seems I'm getting every project except cars. Now I've been asked to make a presentation on means of communication. And it seems some of the Japanese higher-ups will be present. They obviously need to beef up their telephone English. I'm afraid I'm not very good at teaching English.

(In the company auditorium, about fifty people, fade in) ... and so Alexander Graham Bell invented the telephone to force a conversation upon unsuspecting persons in possession of another phone. For those non-native speakers of English in the audience, it consists of the telephone case, the dial and the cradle where the receiver sits.
The receiver is the part that we take from the cradle when we want to make a call. Place the end with the smaller holes to your ear. The other end with the cord will naturally move next to your mouth – like this. If you don't hear a dial tone or another person's voice, you may either have placed the wrong end to your ear or you need to see a doctor. Here we've got a rotary phone. I bet my bottom dollar that you folks from Japan have never seen one in your life. It's a relic of the Empire. However, many pubs and even Ruver have still got some. On this phone you'll notice a dial with ten holes in it. Now, looking at the number you want to dial, stick your finger into the matching hole. Ignore the filth that builds up inside the holes. For each number push the dial clockwise, which means to the right.
A user is referred to as the customer or here in England as the subscriber. The special number you place before the town number is called the dialling code. You have a country or national code for international calls. For example, if you want to talk to O. J. Simpson, you dial 001 for the States and then 310 for O.J.'s home in Brentwood. Whatever you want from a telephone, the various providers have got the answers. Electronic communication now includes car-phones, cellular phones and automatic consoles which enable you to distribute phone-calls all over the company to different extensions. There's even a display, especially for bosses. You can see who's phoning you so you don't have to

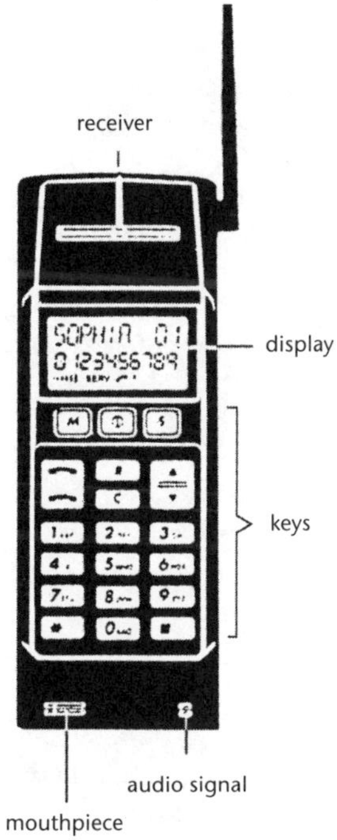

answer if you don't want to. Ha ha! That was only a joke. We all know modern methodology says a communicative boss will find time for any employee who needs him. Well, I think we'd better leave that topic. Connected to the telephone we have the integrated fax, answering machine and telephone, a very practical device ... *(fade out)*
(Fade-in) ... and here at Ruver we should also have internal E-mail and E-mail on the Internet. This means that every employee who has a work-station could communicate with his colleagues. What I also consider very important is a company newspaper or bulletin which appears once a month. By means of this medium we could exchange ideas and inform the large majority of employees. Our many sales personnel in the field would get the latest sales data, prices, conditions etc. from the host, and headquarters would get the latest data from the field. For more urgent matters we should introduce video-conferencing. This is a most interesting ... *(fade out).*

James' diary

God, that was the pits! I thought they were interested in the English terminology around the phone. I'm lucky I've still got a job. Half the audience, especially the Japanese, were asleep. No laughs. I felt like I was at a hanging party, with me being hanged. And then this:
Mr Sinlui (Head of Sales, Asia, in perfect Oxford English): "Thank-you, Mr Kelly, for your lecture in English for foreigners. I believe we did not need it. However, at the end of your lecture you mentioned the idea of distributed software linked to our host. This is interesting. Mr Kelly, to initiate this project will be your responsibility!"

T Working with the text

1. Are these statements true (T) or false (F)?

1. At Ruver they have internal E-mail. ☐
2. In Germany a cellular phone is called a 'handy'. ☐
3. Extensions are the phones in the company. ☐
4. On the dial you key the telephone number in. ☐
5. A host is a main-frame computer. ☐
6. At Ruver the sales force can access the host computer. ☐

2. What James might have said but didn't

Replace the words in italics by those used in the text.

1. The *handset* is the part that we lift when we want to make a call.	The .... is the part that we lift when we want to make a call.
2. The various *telephone companies* have got the solutions.	The various .... have got the ....
3. *With the help* of this medium we can exchange ideas.	.... of this medium we can exchange ideas.
4. I think we'd better leave that *subject.*	Well, I think we'd better leave that ....

H How things work in the phone

Toolbox 1: Conventional transmitting

handset
mouthpiece
coils
microphone
diaphragm
loudspeaker
earpiece
carbon granules
resistance
electromagnet
wires
exchange
sound waves

1. Are you a technical genius? Fill in the words from Toolbox 1

A telephone (**1**) .... has a microphone in the (**2**) .... and a small (**3**) .... in the (**4**) ....
A steady current passes through the (**5**) .... when you pick up the receiver. When you speak, the (**6**) .... enter the microphone and cause a (**7**) .... to vibrate. This alters the (**8**) .... of the (**9**) .... in the microphone and creates a pattern of varying electrical current which matches the pattern of the sound waves. These electrical signals travel through metal (**10**) .... to the telephone (**11**) .... and from there to the receiver of the person at the other end.
The receiver is a small (**12**) .... with a diaphragm mounted to it. The electrical signals travel through the (**13**) .... of the electromagnet. The changing current creates a varying magnetism which causes the (**14**) .... to vibrate. The vibrations produce a series of sound waves like the sound waves that entered the mouthpiece at the other end.

Time for a smile

It was an international computer congress in the USA and the delegates were eating the welcome lunch of the congress. A Japanese gentleman was sitting next to a delegate from Texas. After the Japanese had finished his soup, the Texan asked him 'Likee soupee?' The Japanese nodded. Throughout the meal the American asked such questions as: 'Likee fishee?' or 'Likee drinkee?'
When the meal was finished the chairman introduced the Japanese as the guest speaker of the meeting. The Japanese gentleman rose and gave a witty speech in perfect English. After his speech he returned to his seat and asked the Texan 'Likee speechee?'

Toolbox 2: Modern transmitting

multiplexer
demultiplexer
digital signals
decoder
fibre-optic cable
microwave links
electrical signals

2. Same exercise but things are getting even more complex

In the telephone network, (**1**) .... can be converted into light signals carried along (**2**) .... or radio signals transmitted over radio or (**3**) .... Modern telephone systems convert the (**4**) .... (voltage samples) into digital signals. This is done by a (**5**) ...., which translates the voltage samples into binary code, that is a sequence of ones and zeros. The ones and zeros switch electric pulses on and off. At the destination a (**6**) .... transfers the codes back into voltage samples and the (**7**) .... directs the calls to the correct phone.

3. What is 'it'?

It is the part of the phone you hold. ....
It is a thin metal disk that can vibrate. ....
It is a grain of special coal. ....
It changes sound into electrical current. ....
It encodes digital signals into binary. ....
It changes electrical current into sound. ....
It directs the signals to the right line. ....

Situations

1. What you might see and hear in a public telephone box

1. Nehmen Sie den Hörer ab.	.... the receiver.
2. Werfen Sie das Geld ein.	.... your money.
3. Es sind mindestens 10 Pence erforderlich.	A minimum of 10 p is ....
4. Warten Sie auf das Amtszeichen.	Listen for the ....
5. Wählen Sie die gewünschte Nummer.	Dial the number required.
6. Werfen Sie kein Geld ein. Dieser Anruf ist kostenlos.	Do not insert money. This call is ....
7. Notrufe: Wählen Sie 999.	.... calls: dial 999.
8. Wählen Sie zuerst die internationale Vorwahlnummer, ...	First dial the international ......,
9. dann die regionale Vorwahlnummer ohne die Null, ...	then the area code minus initial '0' ... ,
10. dann die Nummer des Teilnehmers.	then the ....'s number.

2. At a Telecom shop

Five questions, five answers: Client enquires about an ISDN System.

What does ISDN really mean?	It's a change from analogue to digital technology.
Can I connect my old telephone?	You'd better buy a digital phone to get all the advantages.
What are the advantages of ISDN?	You have three different numbers at almost the same price.
Is that all?	No, on your display you can see who the caller is. And you can dial from your computer.
And what are the costs?	The basic charge is more expensive but includes several phones; so it works out cheaper.

Beef up your word power

1. What do you need to assemble to get these parts of the phone?

	....
ISDN system	multiplexer
	....

	....
telephone	handset
	....

	....
microphone	electromagnet
	....

2. Do you know the difference between old and new phones?

Old phones	New phones
They have rotary dials.	They have ....
They have cords.	They are ....
Sound waves are converted into electrical signals.	Sound waves are converted into ....
They travel through coaxial copper cables.	They travel through ....

3. Do you know your fax machine?

scanning device telephone line electrical signals
receiving machine pixels

Fax machines use telephone lines to transmit words and pictures. A **(1)** .... splits up an image or a text into a series of tiny squares. They are called **(2)** .... which are encoded into **(3)** .... These can then be transmitted along the **(4)** .... The reverse takes place in the **(5)** ...., which produces a printed image or text.

Q The quiz for our experts

1. Who invented the telephone?
2. "Mr Watson. Come at once. I want you." Is this a quotation from Sherlock Holmes or were these the first words ever spoken over the phone?
3. Is it true that Thomas Alva Edison had invented the loudspeaker and the microphone before Bell?
4. The telephone signals in optical-fibre cables are produced by a
   A semiconductor laser **B** tiny sparking plug **C** infrared bulb
5. Optical-fibre cables consist of many individual glass fibres. Each fibre is as fine as a human hair. How many telephone calls can a pair of individual fibres carry simultaneously?
   A 1,500 **B** 2,000 **C** 20,000
6. What does ISDN stand for?

R Extra Reading: The fitness phone

Lazy, fat teenagers are frequently the primary telephone users in a family. The inventor Kenji Kawakami has found the solution: the fitness phone. It increases fitness, turns fat into muscle and reduces phone bills. The five-kilo standard model means that most calls will run under three minutes.

Time for a smile: The law of the phone

The telephone will ring when you are having a shower. You will reach it just in time to hear the click of the caller hanging up.

The telephone will ring when you are outside the door, fumbling with your keys.
When you dial the wrong number, you never get a busy signal.

Let's do it crosswise

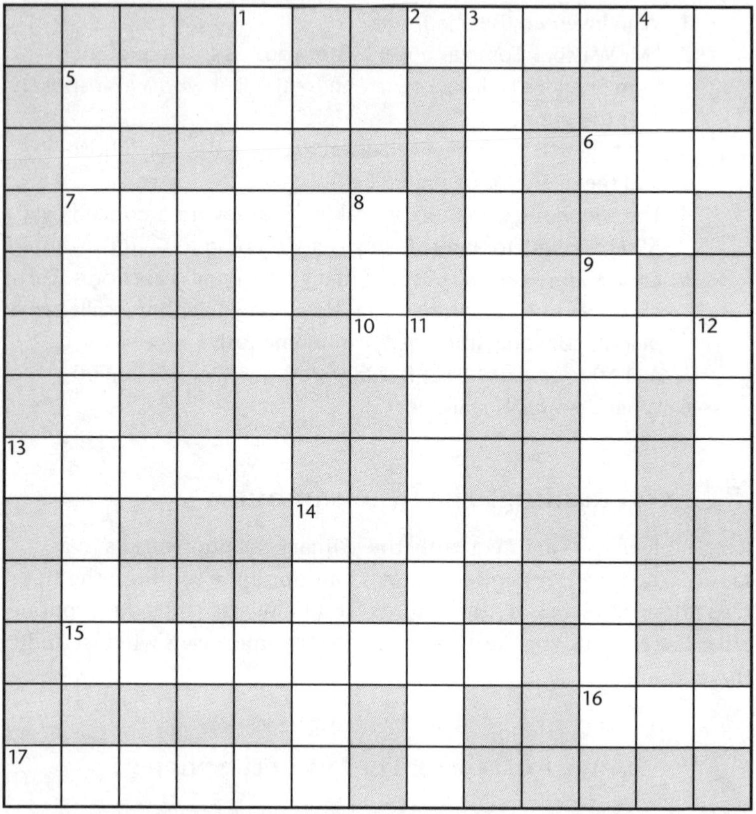

Across	Down
2. the opposite of 'encode'	1. put money into a slot
5. it has two poles	3. opposite of layman
6. it converts text into pixels	4. round disk on a phone
7. English for 'dringend'	5. you speak into it
10. a sequence of different signals	8. turn it and the water will flow
13. funny, intelligent	9. it lies on the cradle
14. it is in a lamp	11. opposite of digital
15. it distributes phone calls	12. movement of the head
16. English for 'Stab'	14. metal plates on a rotor
17. you place the receiver on it	

Ⓥ Vocabulary

advantage	Vorteil
answering machine	Anrufbeantworter
binary code	Binärcode
busy signal	Besetztzeichen
carbon granules	Kohlegranulat
case	Gehäuse des Telefons
cellular phone	Handy
channel	Kanal
clockwise	im Uhrzeigersinn
communicative	mitteilsam, gesprächsbereit
consist of	bestehen aus
console	Display
convenient	bequem, praktisch
convert	umwandeln
cordless	schnurlos
cradle	Telefongabel
current	Strom
destination	Ziel(-ort)
dial	Wählscheibe, wählen
dial tone	Amtszeichen
dialling code	Vorwahlnummer
diaphragm	Membran
disk	Scheibe
distribute	verteilen
ear piece	Hörmuschel
emergency call	Notruf
encode	verschlüsseln, kodieren
exchange; telephone ~	Vermittlung
extension	Nebenstelle
fat	fett
fibre-optic cable	Glasfiberkabel
filth	Schmutz
frequently	häufig
handset	Hörer
hanging party	Hinrichtungskommando

insert money	Geld einwerfen
ISDN	Integrated Services Digital Network
key-in a number	eine Nummer eintippen
match	entsprechen, passen
matching; the ~ hole	das passende Loch
maternity ward	Entbindungsstation
mount (v)	montieren
mouthpiece	Mundstück
native speaker	Muttersprachler
nod	nicken
pattern	Muster, Serie
possession	Besitz
process	ver-, bearbeiten
provider	Anbieter
quotation	Zitat
receiver	Hörer
recently	kürzlich
refer to	verweisen, sich beziehen auf
relic	Überbleibsel
require; be required	erforderlich sein
responsibility	Verantwortung
reverse	Gegenteil
rotary phone	Telefon mit Wählscheibe
sample; voltage ~	Spannungsmuster/-profil
scanning device	Scanner; Abtaster
significant	bedeutend, wichtig
square	Quadrat
supplier	Lieferant
that was the pits	schlimmer ging's nicht mehr
transmit	übertragen
unsuspecting	ahnungslos, arglos
urgent	dringend
urgent matters	dringende Angelegenheiten
various	verschieden(e)
vary	(ab-)wechseln, (sich) verändern
vibrate	schwingen, vibrieren
witty	geistreich

Tools for the presenter

- **OUR STORY: Presenting it all wrong**
- **Working with the text:** Did you get the message?
- **How things work:** In the technical office
- **Situations:** More phrases for meetings and presentations
- **Beef up your word power:** Tools for presentations
- **Extra reading:** The perfect transparency and body language
- **Vocabulary:** Words about presenting

Time for a smile: How to evaluate the presenter

Excellent:	He talks like a god.
Good:	He talks like an angel.
Average:	He talks to himself.
Unsatisfactory:	He argues with himself.
Hopeless:	He loses the argument.

⊛ Presenting it all wrong

James' diary

What do you think of that? The usual story. I keep getting other tasks. And basically I don't know what I'm talking about. Okay, I've got an M.B.A. degree so I'm not completely uninformed. And then the next block-buster. I was called to the Personnel Department.

James Kelly (**J**), *Director Block* (**B**)

B: I've called you here to give you some hints on some things. There are going to be big changes at Ruver in the next months. I'm not permitted even to whisper about them. As far as you're concerned I'm afraid you've got yourself unpopular with some people. You have, however, scored bulls-eyes with some others, especially with our German ... err ... with our Japanese friends. I only have to mention your suggestions regarding Indonesia about the energy from vegetables.

J: Well, thank-you, Mr Block. I know it's a question of ups and downs. But what are you getting at?

B: Look, things may change in the near future. You may be required to become more of a manager, a general trouble-shooter, if you want to stay with the company.

J: I'm beginning to understand.

B: I'm glad. Look, we've certainly a few things to tidy up. Your formal communication skills, are, to put it mildly, weak.

J: Well, I was employed as an engineer. And ...

B: Look, I'll give you a few hints. That presentation you gave on communication: First of all the content wasn't appropriate. Second, you didn't use the basic presentation tools. And third, your body language did not suit the occasion.

J: My body language? Well I'm ready to learn.

B: First things first. You spent half of the time talking about English terminology around phones. Why that? The phone's been with us for years. Why didn't you concentrate on channels of communication, the company's internal E-mail, and the ideas for the future? You didn't even mention our future LAN , the local area network, and what about video-conferences? These are the way forward.

J: I thought they wanted me to up-date their telephone English. It was an unfortunate misunderstanding.

B: One thing's for sure. Talking about dials and telephone receivers is not going to get you far.

J: I get the point.

B: Let's leave that then. What about the technical matter of the presentation? You didn't use the equipment. You had a theatre specially designed for professional presentations. There was a screen for the overhead projector. You press a button, the conference table slides apart and the conference OHP pops up. You didn't need to bring your own. You've got a CAD workstation with the necessary software there. It's standard in this company for transparencies. Why didn't you make use of the company presentation videos? And why the hell didn't you use the laser pointer?

J: I'm afraid my hand was shaking too much for that.

B: We all noticed that when you did your writing or should I say scribbling on the transparency: It was hopelessly overloaded. And the other thing I'd like to mention was your body language.

J: Well, was it all wrong, too?

B: It wasn't all wrong but it wasn't perfect. You didn't seem to be convinced about your subject. Sometimes you were standing with your hands crossed below your belt. Like a priest, we call it the 'pastor' position. You looked like a virgin on her first date.

J: Well, Dr. Block! Did I really look like a virgin?

B: Worse, like a trembling virgin, if I may say so. Not all the time, of course. At the end of your presentation you were standing with both hands in your pockets. This position is actually called the 'double American.' It makes you look like a cowboy.

J: I guess it's my Ohio upbringing.

B: Well, your idea with the distributed software for the field staff saved your bacon, especially with the Japanese.

J: Sorry. I didn't know that. I see I should have gone into everything much more thoroughly.

B: It's in your own interests to sort it out. Perhaps you can satisfy the Japanese with it, but not the Germans.

J: What Germans?

B: Who said anything about Germans? Oh, I almost forgot. I'll lend you my favourite book[1]. It's by a minor German publisher. It's all about presentations. But for heaven's sake don't lose it.

1 *Get Through At Meetings.* Business English für Konferenzen und Präsentationen (rororo sprachen 60265). See also page 128

T Working with the text

1. Are these statements true (T) or false (F)?

1. James Kelly has bought an MBA car. ☐
2. Dr. Block called James to compliment him on his trip to Indonesia. ☐
3. Dr. Block has announced major changes at Ruver. ☐
4. Ruver has a LAN and facilities for video conferences. ☐
5. With a laser pointer you switch the TV on and off. ☐
6. In the 'pastor position' you stand with your hands in your pockets. ☐
7. Obviously there are rumours about a German take-over. ☐

2. What they might have said but didn't

Replace the words in italics by those used in the text

1. What are you *driving* at?	But what are you .... at?
2. You may *have* to become more of a manager.	You may .... to become more of a manager.
3. *You win some, you lose some.*	It's a question .... in Ruver.
4. *The important* things first.	.... things first.
5. We've certainly a few things to *sort out.*	We've certainly a few things to ....
6. I get the *message.*	I get the ....

Time for a smile: Advice for presenters

1. The truth is what you make it.
2. Don't tell lies if you can use statistics.
3. Sincerity sells. If you can fake it, you can't lose.
4. The length of a presentation is in inverse proportion to its value.

H How things work in the technical office

Toolbox for the presenter

A slide projector
B flipchart
C overhead projector
D video recorder

1. Which medium would you choose for ... ?

[ ]	an informal training meeting	3 – 5	participants
[ ]	discussion of projects	5 – 10	participants
[ ]	selling office space	30 – 150	participants
[ ]	introducing new sports equipment	20 – 50	participants

Toolbox for the secretary

stapler
punch
sellotape
liquid paper
drawing pin
paper clip

2. Tools in use: What is the 'it'?

You don't need a dictionary. They are in every office.

It hurts when you sit on it. ....
It is not as safe as a stapler. ....
It is empty when you need it most. ....
It makes holes to fit the folder. ....
It usually sticks to your fingers. ....
It is usually dried up when you need it. ....

Situations

More phrases for meetings and presentations

Put the words from the lists into the right sentence and you won't feel lost in presentations.

1. Check whether you got the message

correct in assuming if I'm wrong
understood you correctly I be right

If I've ...., you are saying that we should give up the whole project?
Would .... in saying that you are against the new model range?
Am I .... that you have a better plan?
Correct me .... , but aren't these figures those from last year?

2. Introduce a question

I wonder Would you mind May I ask

.... where you got these figures from?
.... if I asked you a personal question?
.... if I could ask you what you think of the project ?

3. Ask for explanation

get your last point able to catch
quite follow not quite clear

I'm afraid we weren't quite .... what you were saying.
I'm afraid I didn't .... what you said about ...
I'm afraid I'm .... what you mean by that.
Sorry, I didn't.......................... . Could you go over it again, please?

Beef up your word power

1. What do you get when you assemble the components?

housing	seat
_______ refill	_______ back
cap	stand

keys	microphone
_______ display	_______ loudspeaker
case	tape

2. Lines on your transparency

There could be three different lines on your transparency

a dotted line a dashed line and a continuous line

Write the right name on the line:

This is a .... line. This is a ______________ line.

This is a _ _ _ _ _ _ _ _ _ _ _ line.

3. Shapes for your impact sheets

Can you give them their names? We've helped you a little.

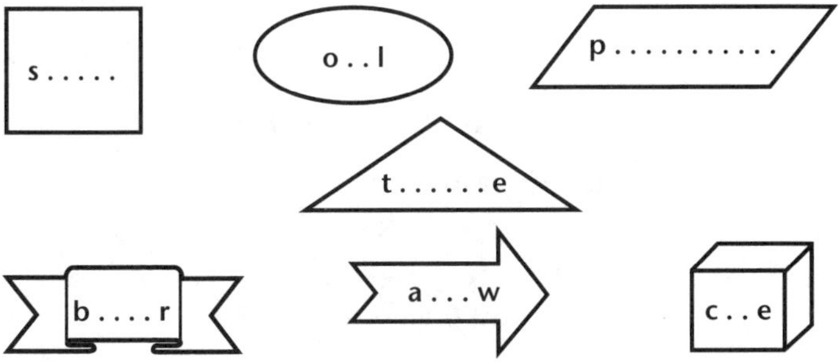

Ⓡ Extra reading: The perfect transparency

You remember? Dr. Block lent James a book about presentations and meetings. James made a few notes for his next presentation.

Rule 1: Don't overload

Don't put more words or graphics on a foil than you would on a T-shirt. If you have to use sentences keep to the number seven. Don't put more than **seven lines** (plus minus two) on the sheet; each line should have between **seven** and **fourteen syllables.**

Rule 2: Not more foils than fingers

Don't use more foils than you have fingers. Seven (plus minus two) is a reasonable number. And you should not speak more than **seven minutes** about one transparency. With **seven foils** your presentation will last about 45 minutes. That's the limit.

Rule 3: The verbal sheet is a killer

Even a transparency with snappy slogans can be a killer. What else can you do than repeat aloud word by word? It's as if you want to teach them how to read And there is another problem. Your audience all listen at the same speed, the speed of your reading, but they all read at different speeds. So you split your audience into several groups.

Rule 4: Impact sheets instead of verbal sheets

An audience is able to take in pictures and listen at the same time to the presenter commenting on them. Use pictures. They make an impact on the audience. At one glance they grasp the message and react emotionally to it.

Food for thought

If you explain so clearly that nobody can misunderstand, somebody will.

To steal ideas from one person is plagiarism; to steal from many is research.

Ⓥ Vocabulary

announce	ankündigen
appropriate	angemessen, geeignet
arrow	Pfeil
assume	annehmen
banner	Band; Banner
block-buster	Knüller; Granate
bull's-eye; hit the ~	ins Schwarze treffen
CAD (Computer Aided Design)	computerunterstütztes Design
catch what someone says	mitbekommen, was jemand sagt
concerned; as far as you're ~	was Sie betrifft
concerned; be ~ about	besorgt sein wegen
content	Inhalt
continuous; ~ line	durchgehende Linie
cube	Würfel
dashed; ~ line	gestrichelte Linie
date	Rendezvous, Verabredung
dictaphone	Diktiergerät
dotted; ~ line	gepunktete Linie
drawing pin	Reißnagel
explanation	Erklärung
fake	nachmachen, simulieren
flop	Mißerfolg, Reinfall
foil	Folie
get far	es weit bringen
go over it again	noch mal wiederholen
grasp the message	die Botschaft verstehen
hint	Hinweis, Tip
impact sheet	Folie, die die Sinne anspricht
length	Länge
local area network	lokales Netz
MBA	Betriebswirtschaftsdiplom
origins	Abstammung, Herkunft
parallelogram	Parallelogramm
pastor position	Priesterhaltung

permit	erlauben, gestatten
point; I get the ~	ich verstehe
pop up; the OHP pops up	der Overheadprojektor fährt aus
proportion; in inverse ~	umgekehrt proportional
refill	Mine (Kugelschreiber)
require; be ~ed to do something	erfordern; etwas tun müssen
rumour	Gerücht
save one's bacon	seine Haut retten
screen	Schirm; Leinwand
scribble	kritzeln
shake	zittern; schütteln
sincerity	Aufrichtigkeit
slide projector	Diaprojektor
snappy slogans	eingängige Slogans
sort out (a problem; something)	ein Problem lösen, etwas in Ordnung bringen
split	spalten
square	Rechteck
staff in the field	Personal im Außendienst
stand	Fuß (Drehstuhl)
ups and downs	ein Auf und Ab
swivel chair	Drehstuhl
syllable	Silbe
transparency	Overheadfolie
task	Aufgabe
tidy up	aufräumen
triangle	Dreieck
unpopular	unbeliebt
value	Wert
vegetables	Gemüse
virgin	Jungfrau
what are you getting at?	worauf wollen Sie hinaus?
whisper	flüstern

The world of copiers

Time for a smile: How to evaluate a troubleshooter

Outstanding:	Walks on water in emergencies.
High satisfactory:	Keeps his head above water in emergencies.
Satisfactory:	Washes with water.
Low satisfactory:	Drinks water.
Unsatisfactory:	Passes water in emergencies.

⊛ Troubleshooting at the advertising department

James' diary

I know I'm always complaining about being given jobs not in my expertise but this time they hit the nail on the head. It all began when I visited the advertising department.

James Kelly (**J**), *Lesley Jones (Head of Advertising)* (**L**)

L: By the way, James, thank-you for dropping by with that information about the styling for the off-road vehicle range. It was the sort of boost we needed.

J: Sounds as if things aren't going too good?

L: At the moment we haven't too much stuff to promote. And when we do have something it takes so long to set it up. You have to make the layout for the advertising brochure, copyright, photos. And all three photocopiers are broken.

J: Well, that's right up my street. Let me have a look. You know, I used to work for a European manufacturer of copiers, Hollerberg in Germany.

L: That would be great. Perhaps you could give us some neutral advice. Maybe we haven't even got the optimal equipment at all. My predecessor ordered all this stuff. But, I'm not sure ...

J: Oh, this is a 2100. That's for general office use. But normally you shouldn't expect to do more than 2,000 copies per month.

L: Well, we consume about 8,000 DIN A 4 sheets a month.

J: Holy cow! So it breaks down a lot. What's wrong at the moment?

L: Several things. It makes dreadful scraping noises when pulling the paper through. Then every half hour or so there's a paper jam. Next is the problem of the toner. The toner cartridge is empty after three days. It consumes more toner than Ruver cars do petrol.

J: That's certainly not normal. Show me how you make a copy.
(Lesley goes to the machine, lifts the rubber flap, places the original on the glass base and presses the green illuminated copy button).
Boy, that's certainly the cause of one of your problems. Didn't you read the instruction manual?

L: Perhaps you've got time for things like that, but we're rushed off our feet. We only skimmed through it.

J: You'd better close the document cover flap or the lid before copying. Otherwise you use ten to twenty times more toner than you need. Regarding the other defects, which symbols show on the display of the copier?

L: These. 1, 2, 3. And 4 is on two of them at the moment.

J: You're right! Number one is paper jam. Can you show me a batch of the paper you use? Ah, recycling, 70 grams. Now, 70 grams is too light. I would suggest that you use a minimum of 80. And symbol 2 is about the transport wheels or runners. They're worn, so they slip over the paper.

L: Because the machine is overloaded. We calculated we spend £15,000 per annum on maintenance and repair.

J: And I think for setting up copies for brochures etc. you need A3 as well as A4 and A5.

L: And when the originals are ready we have to make a minimum of 3,000 for internal company purposes per original.

J: You've got to be joking. You mean you make runs of two or three thousand with these copiers?

L: We certainly do.

J: If I may make a suggestion, get some modern copiers and colour copiers which turn out runs of 5,000 and 10,000 copies in almost offset colour quality. Cost about £ 40,000. Is it the same situation in every department?

L: Must be. Especially in Technical Services, Technical Information Support and General Office Services. It's almost worth having our own printing press.

J: Gosh, no! It takes too long to set up a job. Wouldn't it be better to buy a series of new generation copiers? That would do the trick. And by the way, symbol number 4 means that the image drum is defective. You need a spare part and they're very expensive. But Ruver seems to be able to afford it, from what you've shown me today.

James' diary

And would you believe it? I was ordered to a meeting with Mr Dichasu, Deputy Purchasing Director of Gonda, Europe.

James Kelly (**J**), *Mr Dichasu* (**D**)

D: And so, Mr Kelly, we have been observing your ideas with great interest. Describe the projects you've managed in the last year.

J: Well, it began with the ... *(fade out)*
(fade in). My god, another one sleeping! And that basically was it, Gents. *(coughing)* Hello, are you listening?

D: Thank-you, Mr Kelly. That was an impressive line of projects. And we have one more. It has come to our notice that you have made improvement suggestions for our internal publications systems. We have checked what you have been saying. We would like to offer you the project of purchasing the complete ranges of copy systems for Ruver Europe. We have approval from Ruver top management. Good luck! And your idea to choose from Gondex, one of our subsidiaries, appeals to Gonda very much!

James' diary

Well, my goodness! It's shooting stars with old Kelly. I think that boring Block was just a panic-monger. Of course, I'd have preferred to buy copiers from my former employer, Low Valley Copiers in Hollerberg, but you got to be realistic in the world of technology and business. See you at the top.

T Working with the text

1. Are these statements true (T) or false (F)?

1. The department consumes about 2,000 copies a month. ☐
2. James used to work for a manufacturer of copiers. ☐
3. If the cover flap is open you use ten to twenty times more toner. ☐
4. The department uses recycled paper. ☐
5. James recommends buying a printing press. ☐
6. James suggested buying Gondex copiers. ☐

Time for a smile

If you look like your passport photo you really need the holiday.

Hypochondria is the one disease I haven't got.

H How things work: Getting to know your copier

Toolbox 1

Write the correct number next to the part of the copier. The first one has been done for you.

document glass	2
document cover	
operational panel	
front cover	
handle	
toner/developer cartridge	
paper feed slot cover	
paper feed slot	
paper guide	

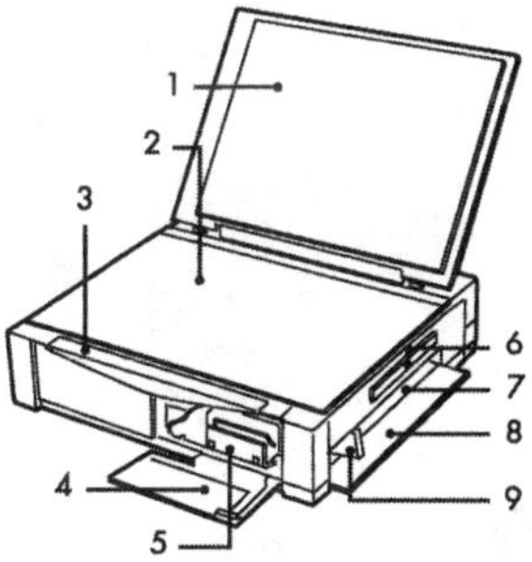

Toolbox 2

release button	
power cord socket	
power switch	

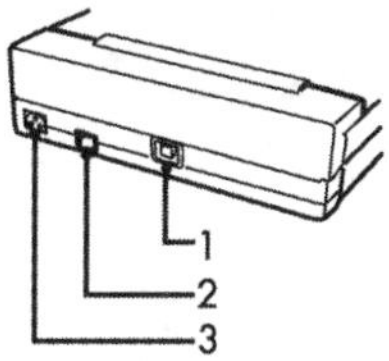

Toolbox 3

photo-conductive drum	
fusing area	
bottom cover	
drum cartridge release lever	

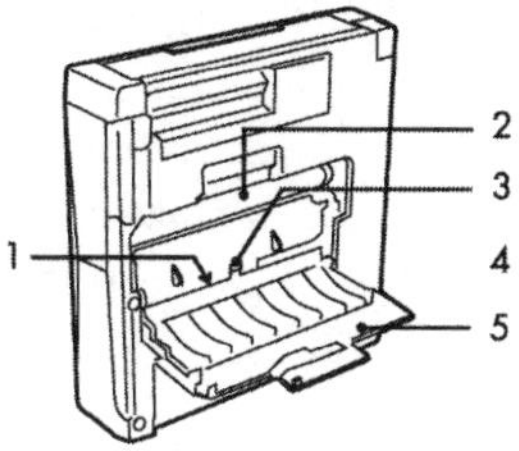

Toolbox 4: The operating panel

key / indicator	nr.	key / indicator	nr.
reduction / enlargement		stop / cancel key	
copy quantity		exposure key	
misfeed indicator		start / ready	

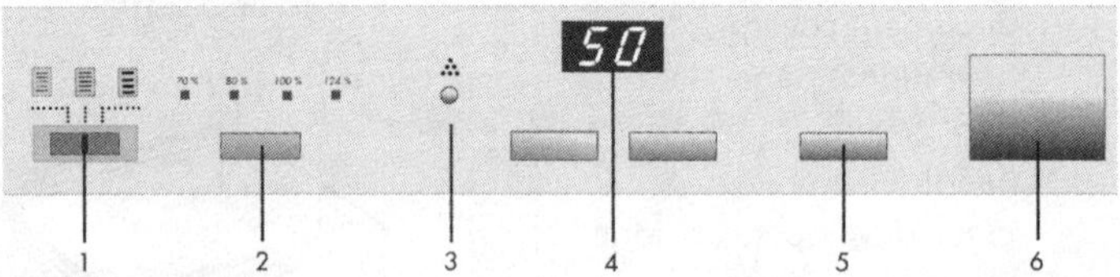

Situations

James has given Lesley Jones some good advice.
Here are more phrases for passing on advice.

don't you		if I may		wouldn't it	
	I would		what about		
it be better		you'd better		why not	

.... think it might be a good idea to plug the copier in first?
.... be wiser to wait unil the photo-conductive drum has cooled down?
.... make a suggestion, always shut the document cover before copying.
Wouldn't .... to buy a series of new generation copiers?
.... use quality paper on this copier.
.... take a more efficient copier?
.... buying the Gondex Copy Miracle?
.... suggest that we put off the decision.

W Beef up your word power

1. What is the 'it'?

It contains black powder. ....
It makes copies smaller. ....
It is a key to blow copies up. ....
It stops the copying process. ....
It blinks when paper jams. ....
It switches the electricity off. ....

2. Making a copy

Here is what you have to do to make a copy.
Choose the right verbs from the list.

switch	turn	remove	open
choose	press	place	insert

1. .... the power switch on.
2. .... the document cover.
3. .... the document on the document glass.
4. .... the exposure for light or dark originals.
5. .... a sheet of paper.
6. .... the start key.
7. .... the copy.
8. .... off the power.

Ⓡ Extra reading: Pages from Gondex handbooks

1. Troubleshooting: Misfeed causes paper jam

What to do when the paper jams in the ...

paper feed slot

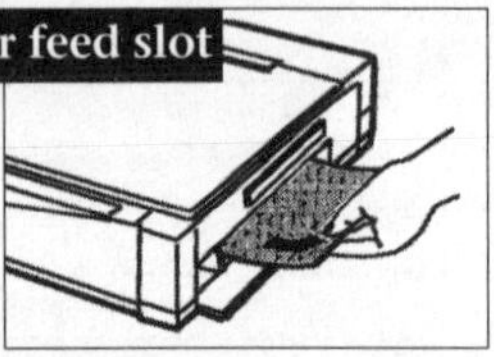

1. Press the release button.
2. Return the document glass to the home position.
3. Remove the document.
4. If all fails, try our new Gondex X3 copier. It is more expensive and but better.

exit area

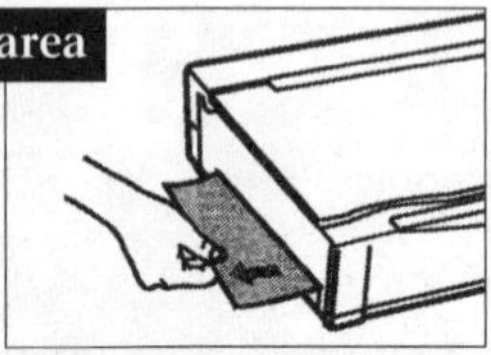

1. Remove the paper by pulling on it.
2. If the paper cannot be removed easily you have used the wrong kind of paper.
3. Carry the copier into the open and burn the paper.

internal area

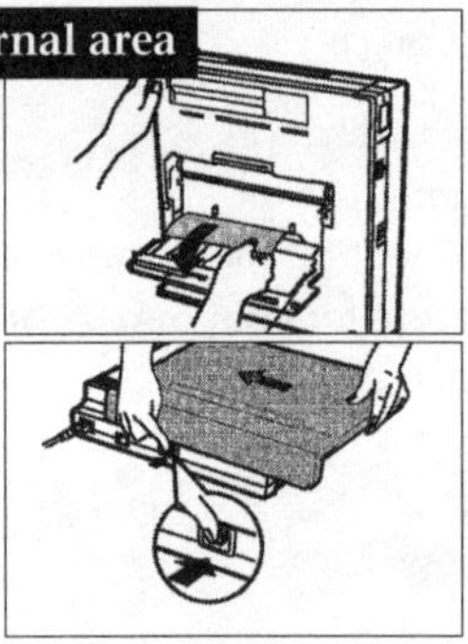

1. Lift the copier by the handle and stand it up with the exit area on the bottom.
2. Push the tab on the bottom cover to the right to open it.
3. Carefully remove the misfed paper.
4. If the paper cannot be pulled out easily, remove all the documents and books within three meters of the copier.
5. Spray the internal area with Gondex Paper Killer. After 20 seconds the paper will dissolve.

And if all fails, go to the copy shop round the corner.

Time for a smile: Don't worry about your product

1. If the product kills 10 per cent of the market, that still leaves you the remaining 100 per cent
2. If you are a big enough company, your mistakes become standards.

2. Troubleshooting: General problems

Our Gondex X 2 and X3 models are very simple to instal. You should assemble the Gondex in its final position, in an area that suits your Gondex. It should not be damp, too hot, humid, poorly ventilated or infested with mice.

The Gondex 3 handy troubleshooting chart

problem	cause	solution
Copier does not copy	You're so forgetful!	Plug it in, you fool!
Copier still doesn't copy	There's something wrong with the copier	Phone Gondex, Virgin Islands
Frequent paper jams	Paper is damp because office is too damp	Use the Gondex paper drier
Copier is noisy	Office is too quiet	Turn up radio
Copies are crumpled	The paper is too dry because office is not	Use the Gondex paper moistener
Copies have dark spots	There are too many mice in the office	Use the Gondex G605 exterminator
Copies are striped	Document may be striped, too	Clean document with Gondex Proper
Copies are spotted, striped and crumpled.	Office is too damp, too hot, poorly ventilated and infested with mice	Change office or use Gondex Copy Shop round the corner

Time for a smile: The law of the copier

1. Any copier can be made to work if you fiddle with it long enough.
2. It works better when you plug it in.
3. If you mess with it long enough it will break.
4. If it breaks, it needed replacing anyway.
5. If all fails, read the instructions.

C Verbs, verbs, verbs – let's do them crosswise

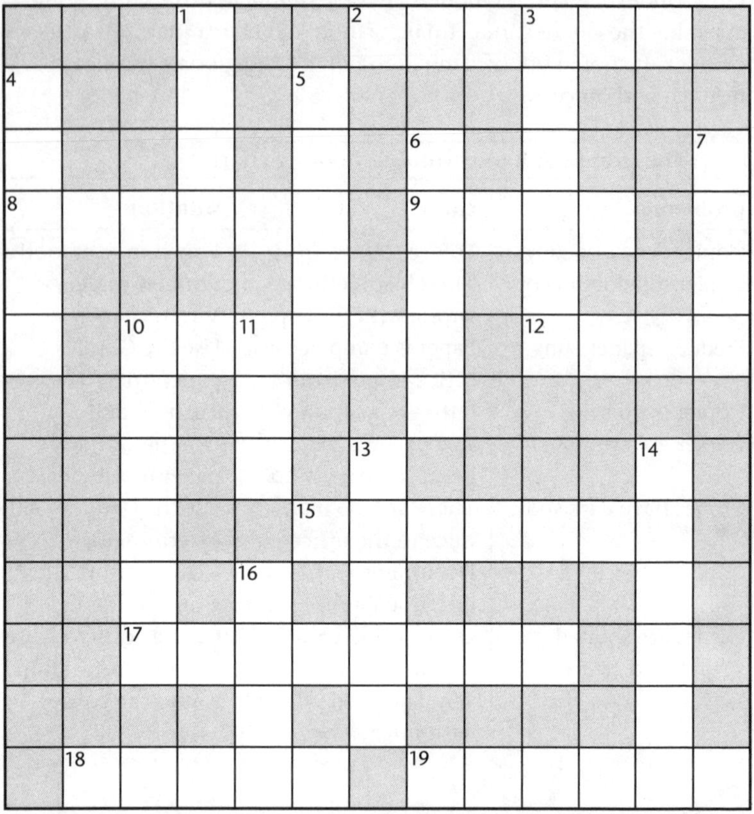

Across

2 give advice
8 these ideas – ? – to me
9 opposite of 'close, shut'
10 always – ? – your seatbelt
12 What a mess! – ? – up the room
15 advertise products
17 give someone good or bad marks
18 René does it when he is nervous
19 stop a job

Remember! They are all verbs.

Down

1 – ? –it out of the box
3 – ? – the paper into the slot
4 – ? – documents on the glass
5 – ? – the board into halves
6 your machines – ? – too noisy
7 you do it on the copier
10 play or mess around with
11 – ? – up a job
13 repair the word: E R U G A
14 opposite of 'insert, place'
16 – ? – a car up

Ⓥ Vocabulary

advice	Rat
advise	raten, beraten
agree with	sich vertragen mit
appeal to someone	jemandem gefallen
appeal to; the ideas ~ to me	die Ideen gefallen mir
argue	streiten, argumentieren
average	durchschnittlich
batch	Charge, Stapel, Stoß
boost	Schub, Anstoß
boost (v.)	fördern, in die Höhe treiben, ankurbeln
bottom cover	untere Abdeckung, Bodendeckel
cancel key	Löschen-Taste
consume	verbrauchen
cover flap	Abdeckung, Bodendeckel
damp	feucht
disease	Krankheit, Seuche
dissolve	auflösen
document cover	Originalabdeckung
document glass	Originalauflage, Glasplatte
dreadful noises	schreckliche Geräusche
drum cartridge release lever	Entriegelungshebel der Trommelkartusche
emergency	Notfall
enlargement	Vergrößerung
evaluate	bewerten, benoten
exit area	Papierauslaß
expertise	Fachkenntnis
expertise; jobs not in my ~	Aufgaben, die nicht in mein Fachgebiet fallen
exposure indicator	Belichtungsanzeige
exposure keys	Belichtungsregler
exterminator	Vernichter
fiddle with	herumspielen, -fummeln mit
front cover	Frontverkleidung

fusing area	Fixierzone
glass base, document glass	Glasplatte
handle (n.)	Griff
hit the nail on the head	den Nagel auf den Kopf treffen
humid	naß, feucht, schwül
illuminated	beleuchtet
image drum	Bildtrommel
improvement suggestions	Verbesserungsvorschläge
infested	verseucht
insert	einführen
instruction manual	Handbuch
internal area	Innenraum
internal company purposes	firmeninterne Zwecke
it's shooting stars with me	mit mir geht es steil nach oben
knock me backwards!	Was du nicht sagst!
lid	Deckel
lose the argument	im Streit unterliegen
maintenance	Wartung
manufacturer	Hersteller
mess with	herumfummeln, herumdoktern an
mice	Mäuse
misfed paper	falsch zugeführtes Papier
misfeed	fehlerhafte Papierzufuhr
misfeed indicator	Papierstauanzeige
moistener	Befeuchter
notice; it has come to our ~	wir haben Kenntnis erhalten
observe	beobachten, verfolgen
operating panel	Bedienungspult
outstanding	hervorragend
paper drier	Papiertrockner
paper feed slot	Papierzufuhr
paper guide	Papierführung
paper jam	Papierstau
pass water	Wasser lassen
photo-conductive drum	Photoleitertrommel
place	auflegen, anlegen

plug in	den Stecker einstecken
poorly ventilated	schlecht belüftet
power cord socket	Netzkabelbuchse
power switch	Netzschalter
predecessor	Vorgänger
promote	(den Absatz) fördern
put off the decision	die Entscheidung verschieben
ready indicator	Bereitschaftsanzeige
reduction	Verkleinerung
release button	Entriegelungsknopf
remove	entfernen
replace	ersetzen
rubber flap, document cover	Gummiklappe
runners	Laufrollen
runs of 2000	Auflagen von 2000
rush; be rushed off one's feet	gehetzt werden
satisfactory	zufriedenstellend
scraping noises	kratzende Geräusche
set something up	etwas einrichten, aufbauen
set up a job	Arbeit, Aufgabe vorbereiten
skim through a manual	ein Handbuch überfliegen
slip (v.)	rutschen
sound (v.)	klingen
spare part	Ersatzteil
style (v.)	gestalten
switch	Schalter
tab	Schieber, Riegel
take long	lange brauchen
that's right up my street	auf dem Gebiet bin ich Fachmann
toner cartridge	Tonerpatrone
toner/developer cartridge	Entwicklereinheit
transport wheels	Transporträder
unsatisfactory	ungenügend

10 Chapter Building your own house

Time for a smile: Mayday, Mayday

The lifeboat was called out to a Japanese yacht in dirty weather. The coastguard, trying to get the yacht's exact location, called it on the radio. "What's your position? What's your position?"
And the answer came, faint but determined, from the skipper: "My position ... ? Well, I'm the new Managing Director of Ruver, a major British car manufacturer."

(*) Taken over by AMW

James' diary

It is autumn 1996. The unexpected has come about. I should have listened to old Block, I guess. On Friday I was a safe employee at Ruver. On Monday morning I became part of AMW, the successful German manufacturer of up-market cars in a more sporty style. AMW had bought the majority interest in Ruver. The news came as a shock. A good friend of mine from Ohio back home, Steve Merrick, was having one of those good English ales with me. The interesting thing about Steve is that he has been working at AMW for six or seven years. He'd got himself a good job as head of a staff department responsible for rationalisation.

James Kelly (**J**), *Steve Merrick* (**S**)

S: And so now that AMW's acquired Ruver there are going to be some changes.

J: Yes, what does it mean for us over here in Britain?

S: The CEO's told us that he intends to bring your quality up to the standard of AMW. That means he'll send a team over to England to work with and replace some of Ruver top management.

J: Aren't the two companies in competition with each other?

S: The CEO insists that the ranges are all in different markets so there won't be any redundancy, but I can't agree with it at all. In the next couple of years they'll have to eliminate all parallel processes, production, design, purchasing etc. And then all areas will probably slim down. Natural wastage will account for some job losses but not all.

J: How will management structure be affected here in Britain?

S: At the moment there's a first-line, a second-line like a division or main department leader, and a department leader. I've heard they'll reduce that to two. Just first-liners like directors who delegate to department leaders. And we want everybody on project contracts or negotiated contracts. The CEO wants to get as far from tariff agreements as he can – more flexibility. For you over here it'll mean more management by objectives and if you don't achieve your goals – well ...

J: Just like back in the States. Will you be involved in the restructuring?

S: You bet I will. I have to identify redundant processes, do out-sourcing to cut costs. Too many things are manufactured in the company that could be done faster and cheaper outside.

J: Well, I see some changes on the way.

S: We're sending our Señor Lupos. He came from the major competitor of your former employer in the States. He's an expert in reducing costs in the area of purchasing and out-sourcing. He'll create the Lupos syndrome at Ruver.

J: Where will I be in all this re-organisation?

S: Listen James, from what you've been telling me, you'd better keep your head down. You got pretty unpopular with some Ruver people and really popular with the Japanese. I advise you to listen carefully to what the Germans say.

J: Why are our eastern friends in AMW's bad books?

S: German top management made a mistake. They thought they could buy out Ruver and everything would stay as it was.

J: I see what you're getting at. The Japanese manufacture the engines.

S: And they stopped it when the Germans got the controlling interest. It's chaotic now. Ruver and AMW have to design everything again. The legal situation is not clear. So if I were you, I'd be very flexible and very co-operative. Good luck.

James' diary

Well, the result of all the reshuffling in the company is that everybody will be required to take a three-week holiday. This will give me some peace and quiet to continue a pioneer tradition that dates back to my Ohio forefathers who used to whack down trees and form them into log cabins. I haven't made up my mind yet whether to build a wooden house or a stone house.

Time for a smile: Why James wants a house of his own

The Ruver guest house is so damp, when James laid a mousetrap in the cellar he caught a herring. However, it is a waterproof house. After a rainfall not a drop leaks out of the cellar. And it is so far from Ruver that the postman sends the post by airmail.

Working with the text

1. Are these statements true (T) or false (F)?

1. Ruver have bought the majority interest in AMW. ☐
2. AMW manufacture up-market, off-road cars. ☐
3. AMW promised not to reduce the number of employees. ☐
4. AMW will negotiate salaries with the trade unions. ☐
5. Those who don't reach the target will be fired. ☐
6. Steve will be involved in the rationalisation process. ☐
7. The Japanese will manufacture the engines for Ruver. ☐
8. James is getting on well with the Japanese. ☐

Food for thought: Business

Buying and selling is essentially antisocial.
Edward Bellamy

If I were a medical men, I should prescribe a holiday to any patient who considered his work important.
Bertrand Russell

What is robbing a bank compared to founding a bank?
Bertolt Brecht

The public be damned! I'm working for my stockholders.
William Henry Vanderbilt

An economist is a guy who would throw you a boomerang for your birthday.
Milton Berle

2. What they might have said but didn't

Replace the words in italics by those used in the text

1. Aren't the two companies *competitors*?	Aren't the two companies .... with each other?
2. Will you be involved in the *re-organisation*?	Will you be involved in the ....?
3. I have to identify *parallel* processes.	I have to identify .... processes.
4. You'd better *keep out of the way*.	You'd better ....
5. Why are our eastern friends *unpopular* with AMW?	Why are our eastern friends ....?
6. I see what you're *driving* at.	I see what you're .... at.

H How things work: Building your own house

1. Why James decides not to build a wooden house[1]

Wood has been the favourite building material for thousands of years because it is one of the few materials that will rot as well as burn. The wood you need for your house is called timber. Since James Kelly is a true Ohio man and an ambitious do-it-yourselfer his first idea was to build a western style wooden house. So James went to a timber yard.

James (J); *Lumberman* (T)

J: Hi. I'd like to buy wood to build a house.
L: What exactly do you need, my boy?
J: All that's necessary for a good solid little house.
L: Well, all you need are footers, joists, beams, rafters, girders, partitions, studs. How many of each?
J: Uh, well, ah, perhaps I'd better go home and recheck my measurements.

1 These invaluable pieces of advice have been adapted from *The Taming of the Screw. Several million homeowners' problems sidestepped* by Dave Barry. Rodale Press, Emmaus, Pennsylvania. The authors cannot recommend this guide warmly enough.

The amateur carpenter's woodbox

strips	Leisten
roof beams	Dachbalken
joists	Deckenbalken
slats	Latten
studs	Pfosten
girders	Trägerbalken
boards	Bretter
rafters	Dachbalken

So James went home, consulted *The Handy Guide for the Homeworker* and decided not to build a wooden house

2. Why James decided not to build a stone house

The stones you normally use for building a house are called bricks. And that's all you need, because, unlike wood, a brick is a brick. Bricks have a big advantage. They neither burn nor rot, they last for thousands of years. The only drawback is that they can fall on you and kill you. Here are few things that might come in handy.
So James bought *The Handy Guide To Building Your Own Brick House.* Here is some invaluable advice on how to avoid the most common mistakes.

1. Get some land first, because most authorities require that your house be built on some sort of land. The best place to build your house is on a slope overlooking a romantic valley. The disadvantage of it is that these sites are very expensive and that you need a lot of money from your bank.
2. The accepted method is to drive two stakes into the ground and attach a string between them to serve as a guide for where your wall will be.
3. Attention! Many so-called professionals will tell you that the string and the first layer of bricks should be level to the horizon. Forget it. In most cases you would have to cut down a lot of valuable trees to be able to see the horizon at all.

4. No! Your string should be level to the ground, that is parallel to your slope. This principle was discovered thousands of years ago by the old Chinese when they built the Great Wall of China to keep out the barbaric Huns and Tartars. If they had been so stupid as to build the wall level to the horizon it would have been so low in some places that the barbarians could have barged right into China. So they built it parallel to the slopes, which stopped the invaders.

Wrong

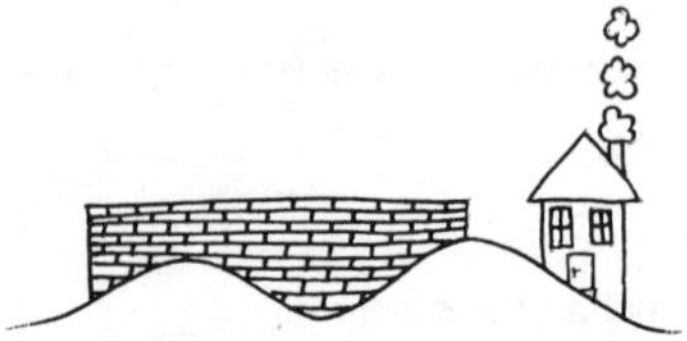

stupid wall and house

Right

intelligent wall and house

5. The rest is mere child's play. You place the bricks layer upon layer until the walls reach the roof. Unfortunately this is where carpentry and woodwork comes in again.
6. Go to a lumberyard and buy girders, beams, rafters, slats, joists and all the rest to make the roof.
   If you're not into wood and carpentry, don't worry. Go to a home centre, buy a prefabricated roof and place it gently, very gently on top of your walls.

Time for a smile: Modern architecture

I asked her if she liked Le Corbusier and she replied "I'd love some – with a little lemon juice in it, if you've got any."

The amateur mason's box

trowel	Spachtel
ladle	Kelle
bricks	Ziegelsteine
mortar	Mörtel
string	Schnur
stake	Pfahl
plaster	Gips
roofing tiles	Dachziegel
spirit level	Wasserwage

3. Why James decides to have the stone house built by experts

The problem with the prefabricated roof is that home centres are always out of everything except artificial Christmas trees and prefabricated roofs for toy houses. "We're expecting a shipment of prefabricated-roofs for slope houses around the middle of next year" was all the support James got from the home centre.

So James decides to have his house built. First, he needs a good architect. However, like good doctors good architects are far and few between. By the way, they have lots of things in common. Doctors spend years studying the architecture of the human body, architects spend years studying the architecture of pyramids and churches. The major difference between a doctor and an architect is, of course, that the first can bury his mistakes whereas the architect can't. And since he has studied pyramids and churches he won't agree with you on what a practical little house should look like. So just tell him how much you're ready to spend on the house, keeping in mind that he will exceed the budget by at least 100 percent.

Next, you need an experienced team of craftsmen, which shouldn't be too difficult, since man had been building solid homes until architects told them how not to do it.

The contractor's team box

craftsmen	their tools	and what to do with them
bricklayer	trowel, ladle	spread mortar on bricks
carpenter	saw, nails, hammer	saw boards & nail them together
tiler	spirit level	make sure the tiles are even
painter	brush, bucket	put paint on the walls
electrician	screwdriver, wires	screw the wires together
plumber	clamps, wrench	clamp the pipes together

4. How to handle the interior decorator

Decorating magazines will tell you that decorating is the easiest part of building your own house. Don't believe it! The best approach is to hire an interior decorator. Interior decorators are people who have spent years studying the principles of colour and shape until they have reached a point where they would rather die than to agree with an ordinary person such as yourself on a matter of taste. So what you have to do is to trick your interior decorator into believing you want the opposite of what you really want. If you want a warm, cosy, intimate look, show the decorator a picture of a Ruver erection shop. If you want a rustic look, show the decorator a picture of the Sistine Chapel. You'll get what you want, and the decorator will think you didn't want it, and everybody will be happy.

The amateur decorator's toolbox

useful things and what to do with them

wallpaper	Tapete	paste it on the wall
glue	Klebstoff	stick things on things with it
trowel	Spachtel	spread paste on with it
cutting board	Zuschneidetisch	cut wallpaper on it
brush	Pinsel	paint surfaces with it
adhesive strip	Klebestreifen	stick things together
paint roller	Malrolle	paint the ceiling with it
bucket	Eimer	keep paint in it

C Let's do it crosswise

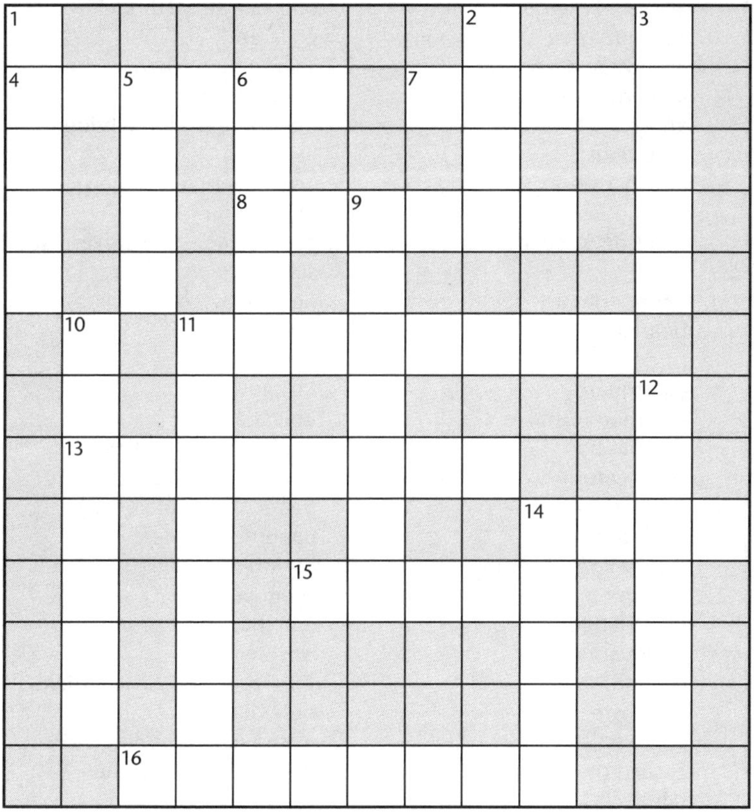

Across

4 repair the word 'EMAUSS'
7 a row of bricks
8 another word for 'provide'
11 English for 'verkünden'
13 location of a ship
14 place where a house is built
15 walls are – ? – to the horizon
16 creating the interior of a house

Some words have been taken from chapter 9

Down

1 very weak
2 a room has four of it
3 stone for walls
5 read very quickly
6 he handles bricks and mortar
9 another word for 'supply'
10 English for 'ersetzen'
11 put components together
12 have the intention to do
14 side of a hill

Ⓥ Vocabulary

accepted method	anerkannte Methode
account for	erklären
achieve one's goal	sein Ziel erreichen
acquire	erwerben, kaufen
affect	betreffen, beeinträchtigen
ambitious	ehrgeizig
attach a string	eine Schnur anbringen
authorities	Behörden
bad books; be in someone's ~ ~	auf jemandes schwarzer Liste stehen
barge right in	hereinplatzen
boards	Bretter
brick	Ziegelstein
coastguard	Küstenwache
common mistakes	übliche Fehler
competition	Konkurrenz, Wettbewerb
competitor	Konkurrent; Rivale
consider	betrachten, halten (für)
contractor	Bauunternehmer
controlling interest	Mehrheitsbeteiligung
craftsmen	Handwerker
determined	entschlossen
disadvantage	Nachteil
economist	Wirtschaftswissenschaftler
eliminate	beseitigen
essential	wesentlich, grundlegend
exceed the budget	das Budget überziehen
faint	schwach
far and few between	dünn gesät, selten
favourite material	Lieblingsmaterial
first-line	Führungsetage
found a bank	eine Bank gründen
girders	Trägerbalken
goal	Ziel
handy	praktisch
home centre	Baumarkt
intend	beabsichtigen

invader	Eindringling
invaluable	unschätzbar
involved ; be ~ in	beteiligt sein an
joist	Deckenbalken
keep one's head down	sich unauffällig benehmen
layer	Lage, Schicht
level / parallel to the ground	parallel zum Boden
lifeboat	Rettungsboot
location	Standort, Position
management by objectives	zielorientiertes Management
mason	Maurer
measurements	Maße
mortar	Mörtel
out-sourcing	extern vergeben, beziehen
paste	Leim
position	Position, Lage; Stellung
prefabricated roof	Fertigdach
prescribe	verschreiben
purchasing	Einkauf
redundancy	überflüssige Arbeitskräfte
redundant processes	überflüssige Arbeitsabläufe
replace	ersetzen
reshuffling	Umstrukturierung
roof beams	Dachbalken
roofing tiles	Dachziegel
rot	verfaulen
site	Bauplatz
slope	Hang, Abhang
stake	Pfahl
stockholder	Aktionäre
string	Schnur
tariff agreements	Tarifvereinbarungen
tiles	Fliesen
timber	Bauholz
timber yard	Bauholzfirma; Sägewerk
unpopular	unbeliebt
up-market cars	Wagen der Oberklasse
valuable	wertvoll
wastage; natural ~	natürliches Ausscheiden
whack down trees	Bäume fällen

11 Chapter

Copiers versus presses

Time for a smile: What is Pi?

Mathematician: Pi is the number expressing the relationship between the circumference of a circle and its diameter.
Physicist: Pi is 3.1415927 plus or minus 0.000000005.
Engineer: Pi is about 3.
Bosewitz: It's pie, not pi, and I like it with apple-sauce.

(*) Printing German style

James' diary

Well, you can bet your booties that things have changed around here. German order and discipline are on the cards. We've all got to be careful.

James Kelly (**J**), *Herr Kurz* (**K**)

(At a big city in the south of Germany famous for its beer)

K: Gentlemen, I declare the meeting open. Thank-you for your time. Speaking of time, Mr Kelly, I should inform you that the meeting began five minutes before your arrival. Here in Germany punctuality is a virtue. Unlike your American ways.

J: I'm afraid I couldn't enter the premises. A problem with the guard at the gate. He wouldn't let me in without the appropriate pass.

K: Herr Schelm, could you please check what happened! Let's get down to business, shall we? The purpose of today's meeting is Mr Kelly's research on re-equipping the European divisions of Ruver for internal and indeed external printing. We consider that you have been very thorough.

J: Well, thank-you, I did my best.

K: But our specialists tell us that under the circumstances it would be preferable to consider printing presses for the quantity and quality of documentation we will need.

J: Copiers are quicker and cheaper for shorter runs.

K: Yes, but our runs will mostly be longer. In addition it is not our policy to use Japanese suppliers. We have a first-class German press manufacturer. They will offer excellent after sales. You have lived in Germany. You know the name of the game, as you English say.

J: Sorry to disappoint you, Gents, but I'm American. But yes, I think I get the message.

K: Could we have your report for purchase suggestions by the first of next month! Is there any other business? ... This is not the case. Well then, I declare the meeting closed. Thank-you, Gentlemen.

James' diary

I left with almost no time and in a hectic. I flew down to Highdelberry, a small university town with a large printing press company. I was to have a meeting with a salesman whose task it was to put me on the right track. No doubt he would try to make it as expensive as possible for me.

James Kelly (**J**), *Herr Sammuk (star salesman)* (**S**)

S: And as I explained there are different sorts of printing like letter-press and colour-copying, but your company's needs as you've described them make me absolutely sure you need offset printing.

J: Well, I'll take your word for it. Who are the specialists for that?

S: My dear Kelly, you are here!

J: I know I'm here. What's that got to do with it?

S: We at Highdelberry Printing Presses are the market leader.

J: Okay. I see. What are your proposals?

S: You have described that you need medium runs, perhaps 10,000 copies or more. And that you need to change your jobs many times a day. Of course, we have large format presses but I'm sure, our Jitto range is the right thing.

J: Well, can you give me a few reasons why you think so?

S: No problem! Minimum sheet size is 360x520 mm. Plenty of space for you. It's easily adjustable, it has many automatic features, high-volume inking unit, a chrome-plated plate-cylinder, line powder-spray device.

J: Hey, hold on, hold on. I can't take all this in. I'm just a poor car mechanic. I need facts, like how many operators does it need? How much of the time can it actually run? And what about training courses? And last but not least is the question of cost.

S: Mr Kelly, I certainly understand your nervousness. I have an idea. Let's look first at this diagram. I'll explain what happens and then we can go to a press which is operating. Look at the diagram: At the top right you can see the paper pile and the infeed ... *(fade out). (fade in)* ... and so that was a description of the process and the paper travel.

J: Why thank you, Mr Sammuk, but actually it doesn't seem that different to colouring copying. Only a lot more complicated to set up.

The Jitto Printing Press

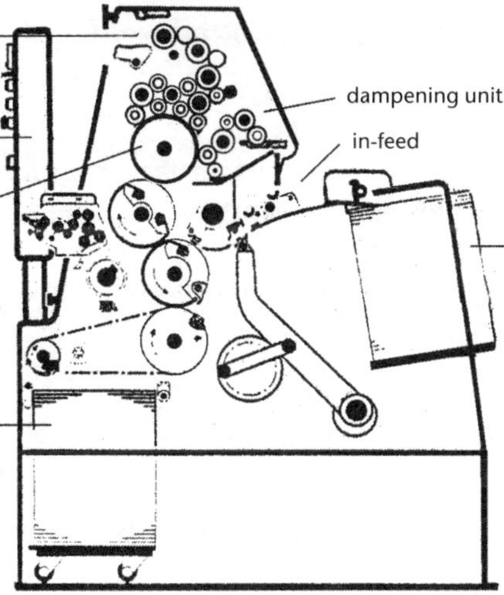

S: I beg your pardon. I must have misheard you. I assure you there is no comparison. The Jitto is high-tech, not high risk and not even high cost.

J: You're the expert, Mr Sammuk. No problem with me as long as it works, regularly, and at the right price.

S: Mr Kelly, I do believe we can discuss all these matters with the help of a good bottle of our German wine and some home-made cooking.

J: Oh, great idea! A bottle of Chianti and steak.

S: I beg your pardon. I must have misheard you again. I assure you there is no comparison ...

J: I know, I know. Your wine is high-tech, not risky and inexpensive. It's certainly a great idea.

(T) Working with the text

1. Are these statements true (T) or false (F)?

1. James is in Munich. ☐
2. James had forgotten his pass. ☐
3. AMW do not want to buy from Japanese suppliers. ☐
4. Printing presses are cheaper for short runs. ☐
5. Herr Sammuk proposes a letterpress. ☐
6. James thinks that the printing press is similar to a colour copier. ☐

2. Words from Chairman Kurz

What they might have said but didn't.
Replace the words in italics by those used in the text.

1. Ladies and gentlemen, *may I open this meeting* now?	Ladies and Gentlemen, I ....
2. The *subject* of our meeting is Mr Kelly's research on ...	The .... of today's meeting is Mr Kelly's research on ...
3. *Now to the matter.*	Let's ....
4. Is there *anything else* to *discuss*?	Is there ....?
5. That *concludes our business for today.*	I .... the ....

Time for a smile

You can tell it's going to be a rotten day ...
... when your boss tells you not to bother taking off your coat.
... when you put on your bra backward and it fits better.

H How things work: Printing presses

1. Printing plates explained in three steps

Toolbox 1

colour scanner	Farbscanner
ink	Farbe
printing plate	Druckplatte
light-sensitive coating	lichtempfindliche Schicht
cylinder	Zylinder

1. Take a coloured picture and place it on a special colour scanner. It produces four separate films of the picture, one for each of the colours yellow, magenta, cyan and black. You get thousands of colours by mixing just these four. Unbelievable!
2. The scanner transforms the illustration into dots or pixels: large dots where the ink must be dense, small dots where the ink must be less dense.
3. When the four films are exposed to light, the picture and the text develop in the light-sensitive coating of the films. Wrap them round a cylinder and your printing plates are ready for action.

2. Offset litho printing explained in four steps

Toolbox 2

lacquer/coat	lackieren
dampen	befeuchten
ink up	Farbe auftragen
press	anpressen

1. Lacquer/coat the letters of the text.
2. Dampen the plate. The lacquered parts reject the water.
3. Ink up the plate. The lacquer accepts the ink.
4. Press the paper against the printing plate. The ink from the inked parts sticks to the paper.

3. The rotary printing press in five confusing steps

However, you needn't be an expert to sort the mess out.

1 The films are	A passed through the press.
2 The printing plates are	B printed one after the other.
3 Sheets of paper are	C wet in the dampening unit.
4 The four colours are	D on one side then on the other.
5 The sheets are first printed	E fitted round printing cylinders.

S Situations: The language of meetings

As you might have noticed, Mr Kurz is a bossy boss and conducts meetings with a rod of iron. One day you might have to chair a meeting yourself. Learn it by doing, that is by doing these exercises.

1. Ten steps to structuring your meeting

item		business		problems
	reason		points	
agenda		consensus		adjourn
	problem		reactions	

1. The .... for our coming together is to ...
2. Let's get down to .... , shall we?
3. Would anybody wish to add an .... to our agenda?
4. There are three .... that should be solved today.
5. The first item on the .... is ...
6. May I ask for .... to Mr Kelly's suggestions?
7. Any further .... before we bring this discussion to an end?
8. Let's now turn to the .... of pollution.
9. To sum up, it seems that we have a .... We're all agreed.
10. Could we .... the meeting to Monday?

2. A guided translation

1. Eröffnung:	.... Gentlemen, I .... the meeting ....
2. Begrüßung:	I'm happy to .... of you to our meeting.
3. Verspätung:	May I .... for the .... that Mr Kelly is late.
4. Ergänzung:	I'd like to .... an .... to the agenda.
5. Zweck:	The .... of today's meeting is ...
6. Tagesordnung:	As you can .... from the .... we must ...
7. Resümee:	If I .... just .... up the discussion.
8. Beenden:	Well, I .... we've .... everything.
9. Schließen:	If .... has anything to .... I'd like to .... the meeting to an ....

W Beef up your word power

1. What is the 'it'?

It separates the colours.	....
It is a coloured liquid.	....
It wets the printing plate.	....
It covers the film.	....
It has a plate wrapped around it.	....
It rejects water.	....

Time for smile: Bumper sticker philosophy

Does the name Pavlov ring a bell?
I'm so poor, I can't even pay attention.
When I grow up, I want to be a Porsche.

A quiz for laymen and experts

1. Without paper no printing. Today paper is made chiefly from fibres produced by trees. Which other raw material was used in Germany before World War II?
   A papyrus **C** palm leaves
   B hemp **D** grass

2. Before paper was invented, people wrote on bamboo in China, clay tablets in Babylon, papyrus in Egypt. Where did people first write on wax tablets?

3. In 105 AD paper was invented. Where?

4. Which people brought the paper to Europe?
   A Phoenecians **C** Arabs
   B Greeks **D** Chinese

5. Where was the first book produced in 868?
   A in China **C** in Egypt
   B in Germany **D** in Italy

6. The first movable types were invented in 1041. Do you know where?
   A in China **C** in India
   B in Germany **D** in England

7. Which material were these first movable types made of?
   A gold **C** wood
   B clay **D** copper

8. When was the first printing press with movable metal type invented? It was so efficient that no significant changes were made until the nineteenth century.
   A 1350 **C** 1450
   B 1550 **D** 1650

9. Who invented the first printing press with movable metal type?

C Let's do it crosswise

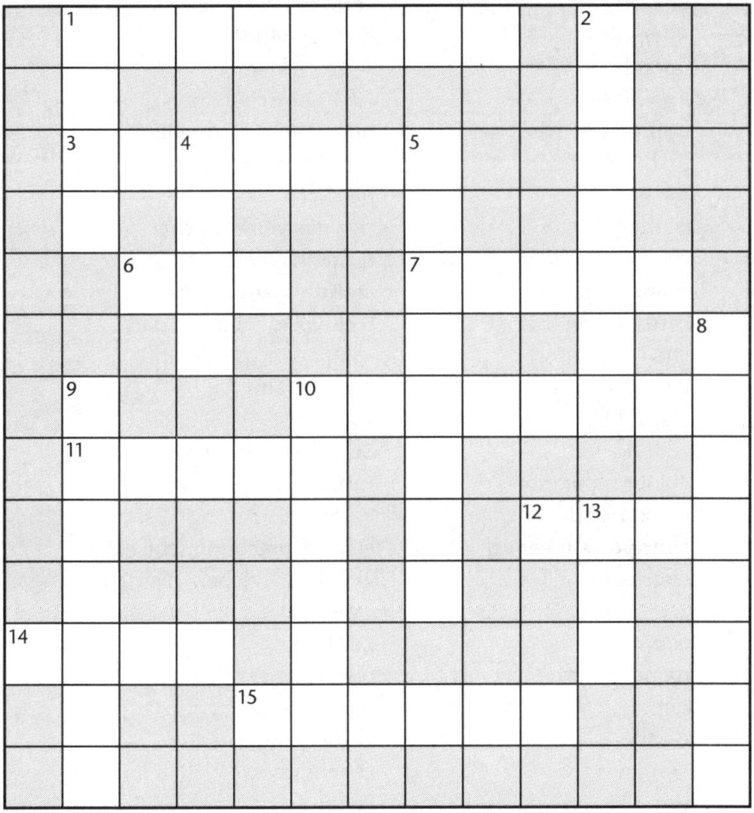

Across

1 English for 'gründlich'
3 you can see it on a bumper
6 another word for 'task'
7 it is fed into the printing press
10 dimension of a circle
11 it's on the agenda
12 it's part of the printing plate
14 plant used to make paper
15 there is no – ? – to be alarmed

Down

1 another word for 'job'
2 theme or topic of a meeting
4 ore to produce steel with
5 repair the word EARERASCH
8 all the buildings of a company
9 not to hear properly
10 English for 'Vorrichtung'
13 coloured liquid

Advice: consult the wordlist

Ⓥ Vocabulary

adjourn	vertagen
after sales	Kundendienst
agenda	Tagesordnung
apologise	(sich) entschuldigen
appropriate pass	hier: gültiger Firmenausweis
assure someone	jemanden (etwas) versichern
bossy	herrisch
bother	sich die Mühe machen
bra	Büstenhalter
bumper sticker	Stoßstangenaufkleber
cards, be on the ~	hier: völlig möglich, daß
chair a meeting	leiten, den Vorsitz führen
circumference	Kreisumfang
clay tablet	Tontafel
coating	Beschichtung, Lack
colour separation	Farbtrennung, -auszug
comparison	Vergleich
conclude a meeting	eine Besprechung schließen
consider	in Betracht ziehen, hier: annehmen, denken
copper	Kupfer
dampen	befeuchten
dense	dicht, verdichtet
device	Vorrichtung, Gerät
diameter	Durchmesser
dot	Punkt
draw the meeting to an end	die Besprechung zum Abschluß bringen
expose to light	belichten
features	Funktionen, Eigenschaften
guard	Wächter
hemp	Hanf
inking unit	Farbwerk
issue	Punkt, Thema

lacquer; varnish	lackieren; Lack
letterpress	Buchdruck; Hochdruck
market leader	Marktführer
matter	Thema; Sache, Materie
message, I've got the ~	ich habe verstanden
mishear	falsch verstehen, mißverstehen
movable types	bewegliche Lettern
name of the game, know the ~ ~ ~	wissen, was gespielt wird
paper travel	Papierlauf
powder-spray	Pudergerät
preferable; it would be ~	es wäre besser
premises	Gelände, Gebäude
proposal	Vorschlag
punctuality	Pünktlichkeit
purchase	Kauf, Einkauf; einkaufen
purpose	Zweck
reason	Grund
re-equip	umrüsten
reject the water	das Wasser abstoßen
research	forschen, erforschen
ring a bell	eine Glocke läuten; an etwas erinnern
rod of iron	Eisenrute, Eisenstange
rotten; a ~ day	ein Unglückstag
runs, shorter ~	kleinere Auflagen
sheet size	Papiergröße, Format
significant	wesentlich, wichtig
stick to the paper	am Papier haften
subject	Thema
task	Aufgabe
thorough	gründlich
track , on the right ~	auf der richtigen Spur
virtue	Tugend
wrap round a cylinder	um einen Zylinder wickeln

12 Chapter

About hardware, software and underwear

Time for a smile: Your computer has the answers

At a computer fair, James was shown a computer that could answer all the questions. He asked it where his father was. It answered "Your father is playing poker in a saloon in Las Vegas." "Wrong," said James. "My father is dead and his body lies in the Baptist Cemetery in Ohio." The salesman was surprised, and hurriedly typed in "Where exactly is Mr Kelly senior?" The computer flashed its answer on the screen: "Mr Kelly senior's body lies in the Baptist Cemetery in Ohio, but this bastard's father is still sitting in that saloon in Las Vegas."

(*) Bits and pieces

James' diary

Well, the bottom line on all that was that I concluded contracts for two million dollars. We've got printing presses everywhere. To tell the truth, I think all the printing could have been done cheaper and quicker outside. Let's see what the future holds. Anyhow, I'm taking Director Block's advice of trying to improve my communication talent. I've got onto a course for EDP and computers and I'm the trainer. I'm looking forward to that especially since it's an individual course and the student is a beautiful Scottish girl! She joined the company a month ago.

Sheila McDray (**S**), *James Kelly* (**J**)

S: *(fade in)* and so I'd be really grateful if you could give me some fundamentals concerning the computer. Dr. Kurz said this new German P.A.S. software was idiot-proof but I ...

J: Well, he should know. If a software is idiot-proof, only an idiot will like using it. Well, I'm sure going to do my best. What is your background?

S: I graduated in arts and applied consumer psychology. And I used to design unisex underwear at the Marks & Spencer research centre and ...

J: Well, you teach me all about underwear, and I'll tell you all about hard and software.

S: You seem to be an expert in all three fields.

J: Actually I'm an engineer, but I guess I've got a thorough hands-on knowledge of computers. A computer is actually a straightforward device that can be built in an afternoon by anyone who has a few simple tools and ten to fifteen pounds of assorted electronic parts. What will you be needing to use a computer for?

S: I'm being trained to take charge of marketing for western Europe. So I'll need the computer and laptop to support my research and work in the field.

J: Okay. I get the picture. Now, I'll give you some background on all the hardware. It's that part of the computer that makes a noise when you throw it out of the window. Let's start with the bits and bytes. You know what they are?

S: I think so. A bit is a piece of computer data. Eight bits make one byte.

J: That's about it. The computer works with electronic circuits which are stored on chips.

S: Those chips are tiny slices of silicon. Each of the chips are covered with millions of circuits

J: Very good. And the chips are mounted on boards called PCBs: printed circuit boards. And the main PCB inside a personal computer is called the motherboard. If you fix smaller PCBs to the motherboard, they're called daughterboards. The motherboard is responsible for seeing that all the daughter boards act properly ... and wear proper unisex underwear. Ha, Ha.

S: *(to herself: This guy seems to have an underwear complex.)* Listen, Mr Kelly, it's beginning to get a bit complicated. I'm afraid ...

J: Don't panic! We don't need to know much more in theory. Buses! What about them?

S: They transport people around towns, don't they?

J: Ha, ha. A little Scottish joke from a little Scottish jokster! Sorry, just another relapse into my American sense of humour.

S: Or lack of sense of humour, I'd say.

J: Hey, take it easy. I didn't mean to tread on your patriotic toes.

S: All right. I'll forgive you this time.

J: The thing is that information flows around the computer in bytes and these bytes are transported along metal paths called buses. So you weren't far off. And there are three types of buses: the data bus, control bus and the address bus.

S: So what's the difference between the three of them?

J: The data bus carries data between the CPU, the central processing unit and the memory or between the CPU and an input or output unit. A control bus ...

S: *(breaking in)* ... carries instructions from the CPU to other parts of the computer.

J: Right. And the address bus carries the numbers or addresses which identify each place in the computer's memory.

S: Well, it's all getting too much for my memory. Do I really need to know all this technical stuff?

J: Sorry if I was getting a bit heavy. You won't be working with one of these big Seamen-Nixgood computers Dr. Kurz wants everybody to use. You'll have to master portables – portable hardware.

S: That sounds better. I'll have a lot to do when I'm travelling.

J: And there's some good software on the market that will cut your workload by 50 per cent.

S: That's great. I'll take two of them.

J: Hm ... I'm afraid it's all getting too much for you. Listen Sheila, what do you say we call it a day. You look tired. I'd say you should give me a chance to make up for my stupid jokes. What about a relaxing glass of white wine at Tati's Bistro just round the corner from my body building club.

S: Well, normally I wouldn't, but since you ask, well okay. I'm into body building, too.

J: Well, we should let our body language speak for us then, ha ha!

Time for a smile

The following day James wrote this computer poem[1]:

My dear Sheila,

If a packet hits a pocket on a socket on a port,
And the bus is interrupted as a very last resort,
And the address of the memory makes your floppy disk abort,
Then the socket packet pocket has an error to report.
If your cursor finds a menu item followed by a dash,
And the double-clicking icon puts your window in the trash,
And your data is corrupted 'cause the index doesn't hash,

Then your situation's hopeless,
And your system's gonna crash!
Yours forever
James

James stole this poem from Doug Mayer *Cyber Jokes*. Harper Collins Publishers

T Working with the text

1. Did you get the message?

1. What is Sheila's background? Did she
   - **A** sell ironware for Marks & Spencer?
   - **B** study arts and psychology at Marks & Spencer's?
   - **C** design underwear for a department store?
   - **D** train to be in charge of marketing for western Europe?
2. What are James' present activities at Ruver?
   - **A** He is building straightforward home-computers.
   - **B** He is training students in electronic data processing.
   - **C** He is improving his communication talent.
   - **D** He is studying computer soft, hard and underwear.
3. Information travels in your computer in
   A bits **B** bites **C** bytes
4. What is a byte?

2. What they might have said but didn't

Replace the words in italics by those used in the text.

1. If a software is *fool-proof,* only a fool will use it.	If a software is ...., only an .... will use it.
2. We don't *have to* know much more in theory.	We don't .... know much more in theory.
3. I *have a degree* in arts and applied psychology.	I .... in arts and applied psychology.
4. A computer is actually a *simple* device	A computer is actually a .... device
5. Listen Sheila, what do you say if *we stop here*?	Listen Sheila, what do you say if ....?

Time for a smile

If architects designed buildings the way programmers wrote programs, then the first woodpecker that came along would destroy civilisation.

H How things work in the computer

Toolbox 1: You and your computer

memory	RAM	keyboard	output unit	input unit
program	ROM	mouse	processor	floppy disk

You have more in common with it than you might imagine. For example, you both have a memory. The only difference between you and your computer is that it has by far the better memory.
You have a short-term memory. It is temporary and to blame when you forget to buy your wife flowers for her birthday. The computer has also a temporary memory, the RAM (Random-Access-Memory), which will only forget, if you forget to save your work.
If you forget both flowers and your wife's birthday, you haven't got long-term memory. Your computer has an excellent one, if it doesn't catch a virus. It's called the Read-Only-Memory (ROM).
You have been programmed by your parents and teachers, your computer by Bill Gates and his programmers. That's why it can add up sums, play chess better and faster than you can.
You take in information with your eyes, nose and ears. A computer has several input units to collect data, such as the keyboard, the mouse or the microphone.
You like your computer because it obeys your instructions without sulking. You take your instructions from your boss or your wife. That's why you do not always like them. Before you execute the instructions you use your brain, at least you ought to. A computer works it out in the processing unit or the processor.
Suppose you have a brilliant idea. Since you are forgetful, you use an agenda or a booklet to write it down. These are your output units. The output units of your computer are the monitor, the printer or the floppy. Floppy and agenda have one thing in common, too. You can't find them when you most need them.
Your brain and your computer obey one universal law:

Input = Output
or in plain English:
Rubbish in = Rubbish out

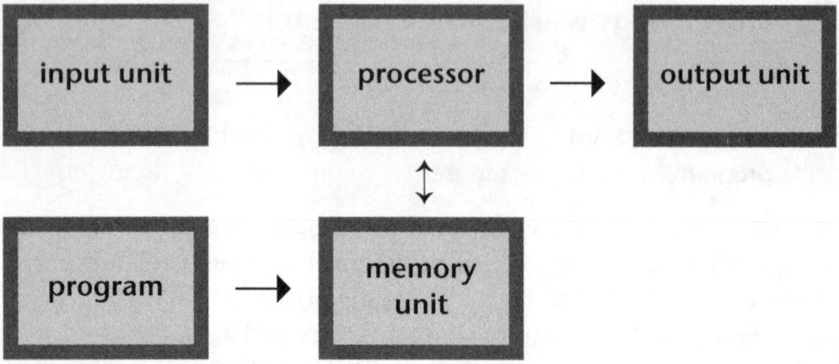

1. Let's put your brain to the test

You have got	What about your computer?
1 short-term memory	....
2 long-term memory	....
3 five senses	....
4 knowledge and skills	....
5 pen, agenda, notepad	....
6 parents and teachers	....
7 grey cells	....

2. The language of the computer

The computer has to change words, pictures, colours and shapes into a language or code that it can think with and remember. It is basically a series of switches which have only two positions: on and off. That is why its language can only consist of two signals: 1 for on and 0 for off. Numbers, pictures and shapes are translated into a series of ones and zeros. This is binary code.

decimal	binary
1	1
2	10
3	11
4	100

Toolbox 2: The memory

floppy disc	Diskette	safety slot	Schreibschutz
guard	Schutzklappe	hard disc	Festplatte
fixing notch	Haltenut	tracking arm	Spurarm
capacity reducer	Kapazitäts-verminderer	read-write heads	Lese-/Schreibköpfe

How does a computer 'remember' things? Most computers have magnetic disks. The on-and-off pulses create a magnetic binary pattern: the directions of the magnetic forces, either north or south.
Your computer uses two different discs: a flexible floppy disc and a hard disc drive or simply hard disc. The floppy is used in a disc drive to store data. You can put it in your pocket and take it to another computer.

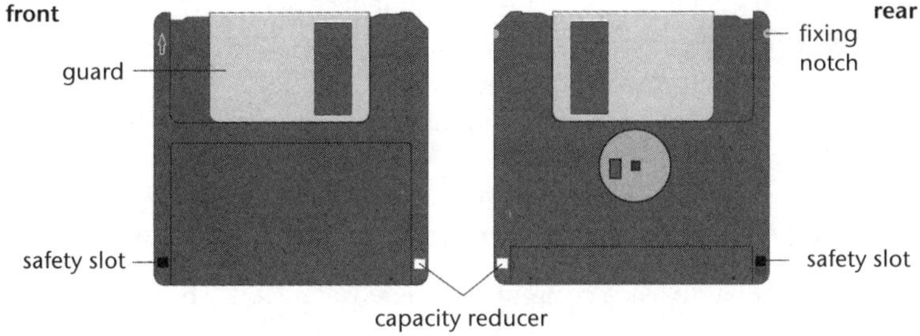

The hard disc has a much greater memory capacity because it consists of several magnetic discs which are in a housing to keep the dust out.

S Situations

Working with your computer

switch on		save		print it out
	open		format	
call up		edit		key in
	retrieve		change	

1. First .... the computer.
2. .... your word processor by clicking on the right icon.
3. .... a new file.
4. .... the name of the file so that you can .... .it again.
5. Write your document and .... it so that it looks good.
6. Correct and .... it, if necessary.
7. Don't forget to .... it.
8. Switch on your printer and ....
9. .... to drive A and make a safety copy on a floppy.

Time for a smile: Your computer has the answers

It was decided to ask a computer if there was a God. The Bible, the Koran, and all the documents relating to any creed, every philosophical argument for and against, were fed in. The computer processed the input for a whole week and came up with the answer: "There is now!"

A computer was installed by the army. An officer asked it, 'How long does a missile take from Moscow to Washington?' '6,000', replied the computer. The officer keyed in the question '6,000 what?' and the computer replied '6,000, Sir!'

W Beef up your word power

1. What do you get when you assemble the components?

	disc
....	guard
	window

	housing
....	screen
	switch

	arms
....	heads
	discs

	circuits
....	silicon base
	photoresist

2. What is the 'it'?

It is the code the computer understands.	....
It is the general name for information.	....
It is an output unit that displays the data.	....
It is the information that comes out on paper.	....
It is the computer's 'short-term' memory.	....
It is the computer's 'long-term' memory.	....
It is a portable magnetic disc.	....
It is a tiny electronic component	....

Time for smile: James' poem to Sheila – part two

When the copy of your floppy's
getting sloppy on the disc,
And the binary instructions
cause unnecessary risk,
Then you have to flash your memory
and you'll want to RAM your ROM.
Quickly turn off the computer
and be sure to tell your mom!

Ⓡ Extra reading: Making your own home computer in six steps

Despite what computer salesmen tell you, a robust home computer is a simple device which anybody whose fingers are not all thumbs can assemble himself or herself.

Step 1: Buy the "Gondex Computer Kit for do-it-yourselfers" at the next Media Market. Wait until your wife and kids are fed up with watching the soap opera.

Step 2: Take a screwdriver and a hammer and remove the back of your Gondex TV set. Weed out everything that is inside, separate metal parts, plastic parts and electronic components – it's the law in Germany – and dump them in the appropriate garbage bin.

Step 3: Spread the contents of the "Gondex Computer Kit" on the carpet and read the 'Gondex Handy Guide to Computer Architecture' until you reach the plan on page 289. The rest is child's play.

Step 4: All you have to do now is to arrange the red, blue, yellow and black parts to exactly the pattern that is shown on the plan.

Step 5: Next you need a soldering iron. Melt solder on all parts until they stick together. Be careful not to melt it on the carpet or the cat. Don't touch the hot parts. Wait until they have cooled off. If you are in a hurry, drop everything into a bathtub of cold water.

Step 6: Once the parts are cool spread Gondex Instant Superglue on all corners and glue everything to the inside of the TV. The glue guarantees a secure a fit. Don't forget to put the back of the TV set back again. Spray the carpet with Gondex Glue Remover.

You are now the happy owner of a home-made home computer. Before you plug it in, check whether your fire insurance is paid up. Switch it on and feed it with bills, checks, receipts and tax records and your new home computer will process data at such a rate that your entire house will be as warm as toast.

C

Let's do it crosswise

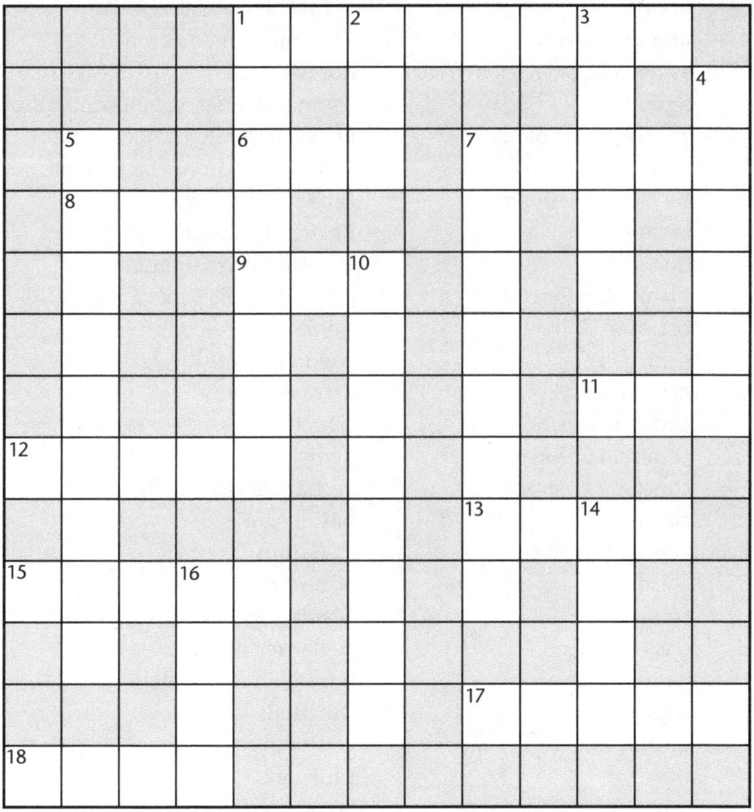

Across

1 what does the P stand for in PC?
6 a computer's long-term memory
7 another word for 'save' memory
8 correct and lay out a document
9 vehicle for data in a computer
11 part of the keyboard
12 you insert a floppy into it
13 another word for 'speed'
15 a computer failure
17 it is measured in mph
18 a document that has been stored

Down

1 a laptop is a – ? –
2 a computer's short-term memory
3 cancel an operation
4 computer brain
5 limited in time
7 all kinds of programs
10 English for 'unterstützen'
14 another word for 'key in'
16 another word for 'store'

Ⓥ Vocabulary

abort	abbrechen, abstürzen
add up sums	Zahlenreihen addieren
address bus	Adressenbus
agenda	Terminkalender; Tagesordnung
applied psychology	angewandte Psychologie
assorted electronic parts	sortierte elektronische Bauteile
be into something	Fan von etwas sein
binary code	Binärcode; Zweiersystem
bit	kleinste Informationseinheit
blame someone	jemandem die Schuld geben
brain	Gehirn
cemetery	Friedhof
central processing unit (CPU)	Zentraleinheit
circuit	Schaltkreis
common ; have in ~ with	gemeinsam haben mit
conclude contracts	Verträge abschließen
consist of	bestehen aus
control bus	Steuerbus
crash	abstürzen
creed	Glaubensbekenntnis
dash/hyphen	Gedankenstrich/Bindestrich
data	Informationen, Daten
data bus	Datenbus
destroy	zerstören
drive	Laufwerk
dust	Staub
edit	bearbeiten, editieren
execute the instructions	Befehle ausführen
fed up; be ~ ~ with	die Nase voll haben von
flash an answer on the screen	eine Antwort auf den Schirm zaubern
glue	Kleber
glue remover	Klebstoffentferner

hash	Kuddelmuddel, Durcheinander
housing	Gehäuse
hurriedly	eilig
integrated circuit	integrierte Schaltung
key in	eintippen
keyboard	Tastatur
long-term memory	Langzeitgedächtnis
make up for something	etwas wiedergutmachen
memory	Gedächtnis; Speicher
menu item	Menüpunkt
missile	Rakete
obey	gehorchen
pattern	Muster
picture; I get the ~	ich bin im Bilde
port	Anschluß
ram	rammen
rate	hier: Geschwindigkeit
receipt	Quittung
reliable	zuverlässig
retrieve	auffinden, abrufen
rubbish	Unsinn; Abfall
save a file	eine Datei speichern
short-term memory	Kurzzeitgedächtnis
slices of silicon	Siliziumscheiben
soap opera	Fernsehserie
socket	Steckdose
solder	Lötzinn
soldering iron	Lötkolben
spirit; the ~ is willing, but the flesh is weak	der Geist ist willig, aber das Fleisch ist schwach
straightforward; a ~ device	ein einfaches Gerät
support	unterstützen
switch	Schalter
tax record	Steuerakte
temporary	zeitlich begrenzt, temporär

thumb	Daumen
trash	Papierkorb
underwear	Unterwäsche
weed out	ausjäten
woodpecker	Baumspecht
yourselfer, do-it-yourselfer	Heimwerker

Computer peripherals made simple

- ✲ **OUR STORY: A working holiday in Scotland**
- T **Working with the text:** Did you get the message?
- H **How things work:** Understanding your laser printer
  Listening to a scanner
- W **Beef up your word power:** Odd man out and
  intelligent guessing
- Q **Our quiz for the experts:** This time it's for the real experts
- C **Let's do it crosswise:** A scanner crossword puzzle
- V **Vocabulary:** Words about accessories

Time for a smile

Sir Harry Lauder, one of Scotland's most famous sons, was once interviewed by a reporter who asked him how he had managed to amass such a large fortune.
"Well, it's a long story," he replied. "and since we have no need of light while I'm telling it, let me blow out the candle."
"I don't think you need to tell me any more," replied the reporter.

(*) A working holiday in Scotland

James' diary

Wow, I never expected to meet a hot cookie like that here in little old England. Oh, I'm sorry. She's Scottish. So we went to my little wine bar. Man, can she drink! You can forget the wine. She wanted lager beer chased by whisky. It's that Celtic thing again. I remember it when I was doing my first job selling sports equipment in Ireland. They were all into drinking. They must be related by alcohol. Anyhow, it was a great evening. I think I'm going to do my bit for intercultural development.
Next week I'm taking three days' vacation and, guess what, so is Sheila. We're going to visit her family in Scotland on a real Scottish farm. It'll be so good to get away from all that computer stuff and the company in general.

Jamie (**J**), *Sheila* (**S**); *Mr McDray* (**D**)

(Sound of driving along in a car)

J: Wow, Sheila, the scenery here is beautiful. It reminds me of home.

S: Jimmy, you know we don't keep a lot of pigs here in Scotland. That's more a task for the English.

J: I didn't mean the pig farming. I was talking about the wonderful mountains and rolling hills. Where did you say your father's farm was?

S: It's just up in the West. Fort William. We'll be there in half an hour.
(Sound of backfiring from engine or exhaust pipe)

J: Damnation! It's not running smooth again. I'm going to get a good German car instead of this English turkey.

S: Calm yourself down, Jamie. In fact, you're driving a good German car now, aren't you? The English certainly aren't good at a lot of things, that's true, but their car-building's not that bad.

J: What sort of farm do you have?

S: Well, it's about 400 acres. Father grows various cereals. He's specialised in feed for various sorts of animals.

J: I guess that's not in fashion just at the moment with all the BSE talk that's going on here in Europe.

S: Up here in Scotland we have nothing to do with their English ways of doing things. We have cattle too, a herd of Angus, and we have a small number of long-haired shaggies.

J: That's a lot of work. How does your dad manage everything?

S: It was a bit hard, but now we're fully electronic. He's got his computer system and all the peripherals. He'll probably want to show you them.

J: Heck, I hope not. We've had enough of that, haven't we?
(Hours later; rattling of dinner plates etc; after dinner discussion.)

D: Well, my boy, at least you're not a sassanach. And you have Gaelic origins.

J: Yes sir, my forefathers are from the Emerald Isle.

D: My boy, since you are here what do you think of the farm? Of course you have much bigger ones back in America.

J: Perhaps we do, sir, but yours is a comfortable, efficient size.

D: Aye lad, but it's a number too big for myself alone. And the lassie is not interested to come in with me on it.

S: Oh dad, that's not quite true. I just don't want to be isolated up here.

D: I know, lassie, you need a man but ...

S: Father could we leave this theme?

J: If I might add something, sir, have you considered electronics?

D: I certainly have, my boy. But I'm at my wits' end with it. Nothing works properly.

J: Probably just some quite logical answer to it all.

D: Jimmy, may I call you Jimmy? Would you come and have a look at it?

J: Jesus, that's no problem If I can help in any way ... *(to himself)* Oh my God. A working holiday. *(sound of door closing)*

D: So here's what I like to call my central tower. In this room you can see everything I use to plan and check my work on the farm.

J: And I see you use a graphics tablet. You can write on the flat pad and it appears on the screen. And look at this scanner and this speech synthesiser. Why do you need to scan anything into a computer?

D: I don't really understand it either. It was all part of the package I bought. The salesman didn't explain it very clearly.

J: I just bet he didn't. You normally wouldn't need more than a hand scanner. That's for small images or sentences. You just drag it or track it across the image. Normally you need a special scan mat or track pad to hold the image still.

D: He didn't give me anything like that.

J: Well, he didn't offer you the hand scanner anyway. You have a full size, A4 colour flat bed scanner as if you were an author of a book or something like that.

D: My dear boy, I have other things, too. Look at this!

J: My God! You're the proud owner of a plotter. If I might ask, what do you use that equipment for?

D: They told me I needed a system, like an output device for my complex drawings and map and all that, for architects plan and so on.

J: But, sir, I hope you'll excuse me for saying so, but you don't need maps and drawings. What for? What you need is a computer with usable software and a printer.

D: Oh, I have the latest printer. It's a very modern laser printer. Here it is:

J: Wow! It's true, it's state of the art. But did you really need that?

S: Pops probably didn't. And it doesn't seem to work, either. James, could you spend a bit of time with Pops?

J: Well, I mean I'll do my best but actually I'm only a motor mechanic. I'm no hot-shot on all this.

S: Jimmy, for me you're hitting the target every time!

J: Oh Wow, I mean ...

Ⓣ Working with the text

1. Are these statements true (T) or false (F)?

1. They are driving in an AMW car. ☐
2. Mr McDray is specialised in breeding various sorts of animals. ☐
3. James has Irish forefathers. ☐
4. Sheila is looking for a husband to take over the farm. ☐
5. Mr McDray needs the peripherals to make maps. ☐
6. A plotter is a sophisticated printer for graphics. ☐

H How things work in laser printers and scanners

Understanding your laser printer in four steps

Step one: A laser fires a beam of light at a spinning mirror. Another mirror and a set of lenses then focus the beam onto a drum, which carries a negative charge.

Step two: The electric charge disappears where the beam strikes the metal surface of the drum. The parts that remain charged represent the dark parts of the document. Signals from the computer turn the beam on and off, building up an electrical image.

Step three: Positive charged particles of toner powder are then applied to the drum. The charged parts attract the powder, which is then transferred to the paper.

Step four: A heater in the fusing system seals the powder to the paper. The image is transferred onto the paper as in a photocopier.

The word 'laser' actually came from the letters that start each word in this phrase: **L**ight **A**mplification by **S**timulated **E**missions of **R**adiation. What it means is that light can be amplified, or made more powerful, by giving it a boost with electricity. If light is amplified enough, it can melt metal. Other lasers don't cut, like the laser in a compact disk player. They just measure the depth of a tiny hole under the clear plastic of the CD, and the measurement tells the machine what sound to produce.

Toolbox: Laser printer

1	paper	Papier
2	toner cartridge	Tonerpatrone
3	laser beam	Laserstrahl
4	spinning mirror	Rotationsspiegel
5	metal drum	Metalltrommel
6	electrical charger	elektrische Ladungen
7	fusing system	Fixiersystem

Now try to put the numbers from the toolbox next to the arrows

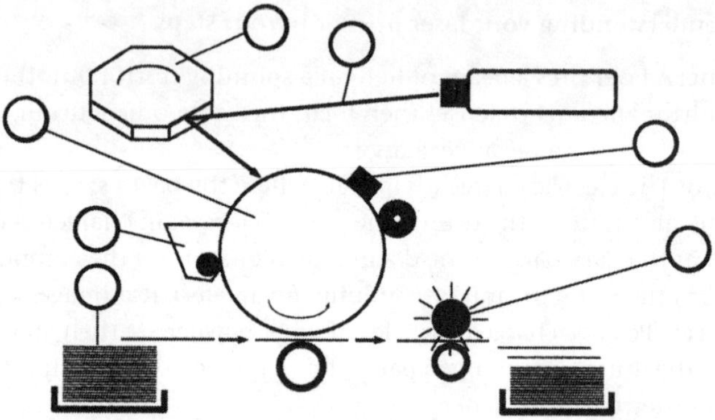

Listening to a flatbed scanner

I'm connected to your computer and I'm intelligent. I can read a text and I can speak it, too. I'm the best friend of blind people, because I can tell them all the stories and letters they cannot read themselves. Let me explain to you how I do it.

I am owned by René B. He puts all the letters from his girlfriend on the glass of my flatbed scanner. My little scanning camera moves along the lines and takes a picture of each single letter. Let's take the "E" in the most frequent sentence: "René, I love you".

What a lousy text! Three different shapes! That's a problem for my OCR (Optical Character Recognition) software. First I convert each letter into electronic signals, also known as ASCII. Then I have to look into my dictionaries of weird words and letters that are stored in my memory. Let me see. Ah yes it must be "love". No problem. But what about this strange **é** in "René". The word is not in my dictionaries. I'll check in my collection of fonts. I got it. It must be this French **é**. I'll store all these funny **Es** in my memory in case I get more love letters from this girl to read for René. Here is an overview of what I've done so far.

René ➔ ?	82 01010010 69 01000101 78 01001110 82 01010010	E e R n è é N ê
I scan abnormal "René"	I convert René into ASCII	I check in my dictionary of fonts and rules

Now, René is not really blind in the ordinary sense of the word, but he is too lazy to read the letter himself. He wants *me* to read the letter to him. So I first convert the smallest meaningful elements of each written word, the graphemes, into the smallest acoustic signals, the phonemes – electronically of course. Then I analyse the punctuation, and break each sentence down into its units to get the stress of the words and the rhythm of the sentence right. How? Well, my programmer has given me all the phonetic rules about how to form syllables – electronically of course. The rest is mere child's play. I send my signals to a voice synthesiser. It is less intelligent than me, but it has been taught all the rules to convert my phonemes into sound and to add pitch, stress and the duration of the vowels. A simple speech synthesiser and loudspeaker do the rest. And that's how René can put a cassette into the cassette player of his Saab and enjoy the sweet words of his beloved while driving to his favourite pub round the corner.

Time for a smile

The Gondex Kit includes a free translation software which is not 100 per cent reliable. I had to translate the phrase "the spirit is willing, but the flesh is weak" into Russian. Just to check I made the computer retranslate it into English. The result was: "the vodka's all right, but the meat is lousy."

Beef up your word power

1. Odd man out

Which words do not belong into the column? Why?

scanner printer computer laser	ASCII phoneme grapheme syllable	heater toner cartridge laser beam spinning mirror	spinning mirror metal drum electrical charges fusing system
translate arrange convert transfer	control check verify correct	store save file retrieve	amplify the beam focus the beam convert into ASCII build up an image

2. Intelligent guessing. What is the 'it'?

It converts computer signals into sound. ....

It focuses beams of light. ....

It is a computer code. ....

It recognises shapes. ....

It is a component that seals toner to paper. ....

It is a super printer for graphics. ....

Food for thought

Experts at the Pentagon were scratching their heads about what to do next. Their most complicated code had been cracked by seven teenagers using ordinary computers.

Our quiz for super experts

1. What does ASCII mean?
2. What do the letters OCR stand for?
3. Where does the word laser come from?
4. What is the device called that bends light rays to form an image?
5. Which Roman Emperor was one of the first people to use a lens when he watched performances in the arena?
6. All rays of ordinary light have the same wavelength. Is this true or false?
7. This is a tricky question. Be careful! A kilobyte is how many bytes?
   A 1000 bytes **B** 1024 bytes **C** 1008 bytes
8. What is a 'cog'?
   A husband of a hen **B** policeman **C** toothed gear wheel
9. What is the point called at which rays or waves meet?
10. Which device splits up the image into a series of lines and converts the various levels of brightness into electrical signals?

Time for a smile

What did the lion say that swallowed a computer?
"I couldn't eat another byte."

A computer is only as stupid as its owner.
Mike Knowles.

I can't understand the fuss about computer pornography.
Surely everybody has seen a computer without clothes on.
René Bosewitz

Computers come in two varieties: the prototype and the obsolete.

C Let's do it crosswise

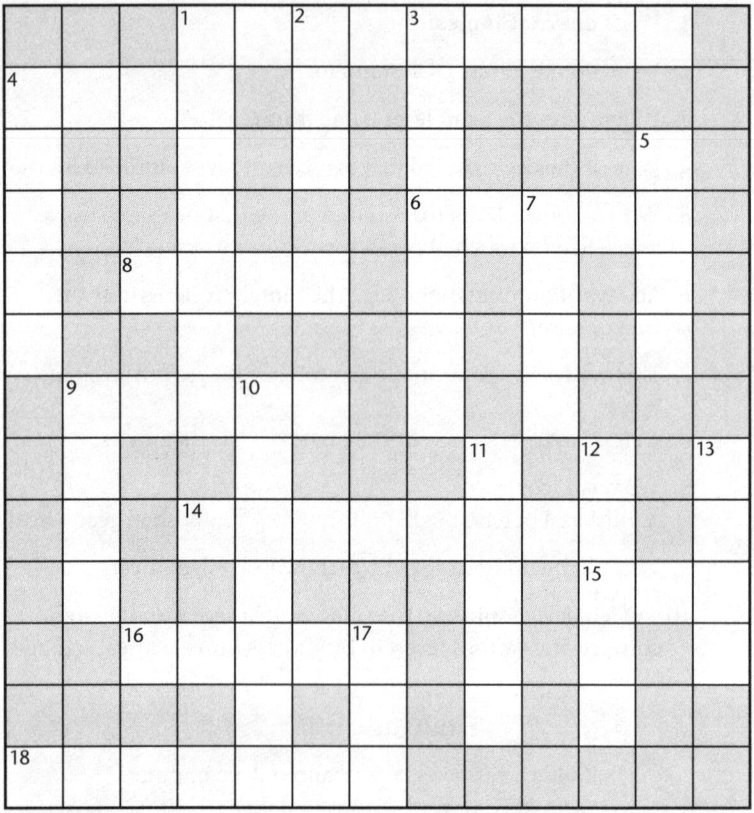

Across

1 what does C stand for in OCR
4 we measure it
6 light to cut and measure with
8 opposite of 'hardworking'
9 Scottish for 'girl'
14 English for 'häufig'
15 scanner software
16 level of your voice: p _ _ _ h
18 best publisher of paperbacks

Down

1 farm products: c _ _ _ _ _ s
2 English for 'Kunst'
3 make bigger or stronger
4 Complete: – ? – of the art
5 part of a printing press
7 conserve with laquer
9 part of a camera
10 s _ _ _ _ _ h one's head
11 you measure it: d _ _ _ h
12 another word for 'save' a file
13 Roman emperor
17 we all try to – ? – the target

Ⓥ Vocabulary

amass	anhäufen
amplification	Verstärkung
apply	aufbringen, auftragen, anwenden
backfiring	Fehlzündung
beam of light	Lichtstrahl
boost; give a ~	verstärken
break down	unterteilen
cereals	Getreide
charge	(elektrische) Ladung
chase; beer ~d by whisky	Bier mit Whisky hinuntergespült
consider	denken an, in Betracht ziehen
convert	verwandeln
crack a code	einen Code knacken
depth	Tiefe
drag across a picture	über ein Bild ziehen
drum	Trommel
duration of a vowel	Länge eines Vokals
exhaust pipe	Auspuff
fashion	Mode
flatbed scanner	Flachbettscanner
font	Type, Schriftart
frequent	häufig
graphics tablet	Graphiktablett
hit the target	ins Ziel treffen
hot cookie	tolles Mädchen
hot-shot	Experte
instead of	anstatt
intercultural development	Entwicklung interkultureller Beziehungen
lassie (Scottish)	Mädchen
lazy	faul
letter	Brief; Buchstabe
measurement	Maß
origins	Herkunft, Abstammung

peripherals	Peripheriegeräte
phonetic rules	Lautgesetze
pitch	Stimmlage
plotter	Kurvenzeichner
pops	Pappi
radiation	Strahlung
related	verwandt
remain charged	elektrisch geladen bleiben
run smooth	ruhig, rund laufen
sassenach (Scottish)	Engländer, Flachländer
scratch one's head	sich hinterm Kopf kratzen
seal	versiegeln, konservieren
set of lenses	ein Satz optischer Linsen
shaggy	zottig, rauhaarig, struppig
size	Größe
state of the art	Stand der Technik
track across	ziehen über
track pad	Trackpad (berührungsempfindliches Display)
variety	hier: Ausführung, Modell
voice	Stimme
weird	sonderbar
wits ; be at one's ~ end	nicht mehr weiter wissen

Auto-pilots for the road

Time for a smile: Unbelievable but true

In 1993 London Transport presented the first prototype of the driverless bus in Nottingham to the public. Journalists, camera men and local politicians were waiting at the computerised bus-stop to board the driverless bus, which arrived absolutely on schedule. There was just one problem – the driverless bus did not stop.

(*) Ruvers with a brain

James' diary

Next time I'll think twice before going on vacation. I didn't get any rest at all what with running around after Sheila's father. In the end we took most of his electronic equipment back to the shop and got it exchanged. Now he's the proud possessor of a small system with an easy calculation software. Sheila and I are getting on like a house on fire. Work is quite another matter. The parent company is doing various management audits. They're trying to rationalise everything just as Steve said they would. Nobody's sure to have a job next month. I get one strange project after another. Most of them are taken out of my hands before I can finish them. Last Friday was a case in question.

At a meeting. James Kelly (**J**), *Herr Kopp* (**K**) *head of European Future Technology from AMW in Germany.*

K: And so, Mr Kelly, we wish to introduce a traffic navigation system. Of course it will not be standard in all our cars.

J: Well, I've noticed that most things are not standard in German automobiles anyway, so it'll fit the philosophy.

K: My dear Kelly, we are not producing cars to American specifications. Quality has its price, you know. Now to work. We must keep to our schedules. What do you know about navigation?

J: Not much I guess.

K: As I imagined. But we have done some research. Let me explain! Already in England millions of pounds have been invested in schemes for using driverless buses. They are called the Lock-On Overland Facility, LOOF for short.

J: What's the technology behind their scheme?

K: Ha ha! The curiosity of an engineer. Even an American engineer!

J: Hey, now just a minute. Cynicism is uncalled for here.

K: No offence intended.

J: Well, there was certainly some offence felt. I ...

K: *(interrupting)* My dear James, I may call you James, where is your American sense of humour that we have all heard about?

J: I guess I seem to be losing it, when personally attacked.

K: Onward! London Transport described it as a "computerised tram". The idea is that it would follow a magnetic track on the road, marked by flashing lights to alert other traffic. It would be linked to a master control panel at London Transport headquarters.

J: And how would this bus 'see' where it was going?

K: There are cameras at the front and rear to scan the road. The pictures are then relayed to the controller who monitors eight buses. He can stop the bus at any time, of course. Come to my office in the afternoon and I'll explain it to you in more detail on the big posters London Transport has sent us *(see page 164, 166)*.

J: It seems to be a great idea.

K: Yes, the buses are fitted with transmitters. The transmitters are activated when the bus passes beacons located underground at key points.

J: So you want me to go into London, check out the technology and find out how it can be adapted to our Ruvers. Great idea!

K: No, Herr Kelly. This is all too primitive. We must be at the state of the art. I have booked you a ticket to Japan. That's where it's all going on.

J: What are the Japanese up to then?

K: We have done our research.

J: I'm sure you have.

K: You can bet your last cent. They use satellites. They have a device that can identify the position of your car to an accuracy of fifty metres on a map stored on a laser disc. The system gets its information from the US government's GPS.

J: GPS? Sounds like the general postal service.

K: Ha, thank-goodness for our research. It means Global Positioning System. From your military. What would we all be without our international American policemen? This GPS is a network of satellites in six orbits.

J: I see. So with the help of these satellites, a receiver located in the car could receive data about its location – geographically so to speak. It could make corrections for the local topography, for hills, rivers and so on.

K: And you have a display on your dashboard. And it's accurate to fifty metres.

J: Why not do it here in Europe?

K: First, in Germany, we have been sleeping with regard to those things. And the rest of Europe is not reliable in these matters either.

J: Well, I noticed there are quite a few things going on in England. I think they ...

K: *(interrupting)* Nonsense. The Japanese are our only real competitors. The second point is "vitamin B".

J: Problems with your health?

K: Nonsense. I do not experience problems, only solutions.

J: So perhaps you're experiencing some solutions with your health? With the help of vitamin A or B?

K: You misunderstand. I mean old boys.

J: Sorry, I didn't catch that.

K: Your saying. Old boys' network, vitamin B. Using connections.

J: Oh, I get you.

K: You are the method. It has come to my notice that you were very warm and cosy with our Eastern friends. You were the teacher's pet, as it were. We have our research, you know.

J: Well, congratulations on that, Herr Kopp.

K: To prepare yourself, please look at this transparency.

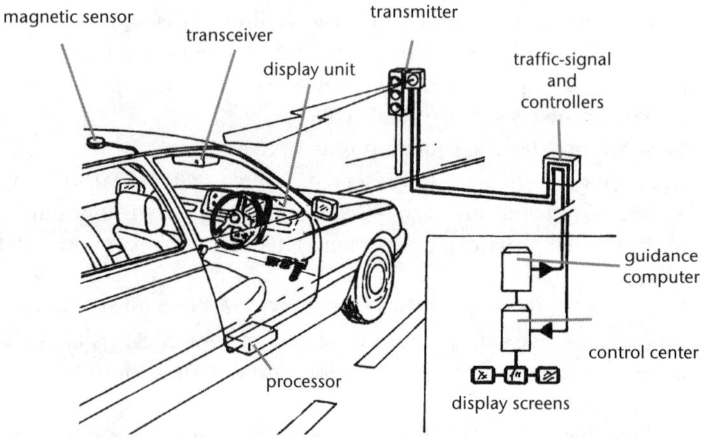

T Working with the text

1. Are these statements true (T) or false (F)?

1. Ruver cars will be fitted with LOOF. ☐
2. The GPS has been developed by Japanese military. ☐
3. Making use of the old boys' network means using connections. ☐
4. James is sent to Japan because he has connections there. ☐
5. The computerised tram gets its signals from satellites. ☐
6. Mr Kopp wants to develop a driverless Ruver car. ☐
7. The Japanese are developing a navigation system on the basis of GPS. ☐

2. What they might have said but didn't

Replace the words in italics by those used in the text

1. We *mustn't fall behind* our schedules.	We .... our schedules.
2. There are cameras at the front and *the back.*	There are cameras at the front and ....
3. The pictures are *transmitted* to the controller.	The pictures are .... to the controller.
4. We must be *up to date.*	We must be ....
5. The Japanese are our only real *rivals.*	The Japanese are our only real ....
6. Quality *is expensive.*	Quality ....

Time for a smile: Graffiti at a London bus-stop

God made things that creep and crawl
But London Transport beats them all.

H How things work

Tool box 1: The electronic ticket machine

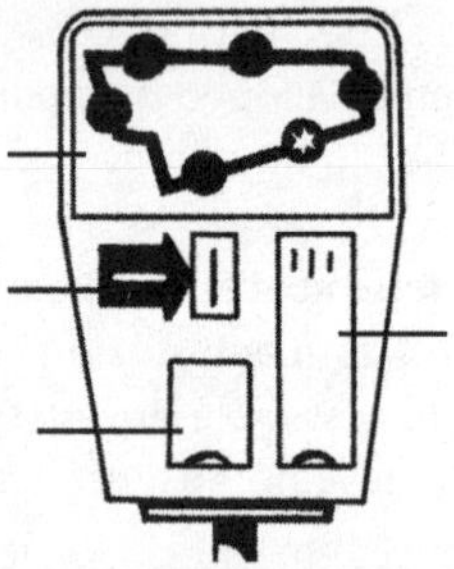

1. Tools in use: At the electronic bus-stop – Use the right terms

You are waiting at an electronic bus stop. The first thing you want to know is how long you have to wait. You take a look at the (**1**) .... which shows you the exact position of the bus and the time you have to wait for it. There are, of course, no ticket collectors like on the traditional buses. How do you know how much money to insert into the automatic (**2**) .... ? Well, press one of the stops displayed on the (**3**) .... and the fare will flash up on the display. You haven't got any small change? No problem. The (**4**) .... accepts pound notes and you can even pay with plastic money. Insert your Visa Card into the (**5**) .... on the left under the electronic map. Don't forget to select the right bus stop or you will be in real trouble when you try to leave the bus. Lift the (**6**) .... over the (**7**) .... and take the ticket out.

By the way, there is only one little hitch. You can't dodge the fare. The electronic turnstiles at the exit doors will be blocked, if you try to get off without a ticket or with one that has not the exact distance programmed into it. Not only will all the other passengers know that you are a dodger, but, what is worse, you will have to stay on the bus and drive round and round the town until, late at night, the driverless bus turns in at the depot where the police will already be waiting for you.

Toolbox 2: The driverless bus

metal granulate	Metallgranulat
entry and exit doors	Einstieg und Ausstieg
flashing units	Signaleinheit
skirt	Seitenverkleidung
magnetic sensors	Magnetfühler
remote control	Fernsteuerung
control point	Überwachungszentrale

2. Tools in use: Understanding the driverless bus

Put the numbers in the picture next to the sentences.

A. Metal granulates are fitted into the bus lane. They form an electronic track (no ...).

B. The flashing units on the track signal to other drivers on the road (no ...).

C. There are front and rear cameras that transmit the pictures to the monitors at the control point (no ...).

D. The entry and exit doors give an acoustic warning before closing (no ...).

E. The skirts have built-in sensors so that the bus can avoid obstacles, for example, when a pedestrian is standing too close to the road (no ...).

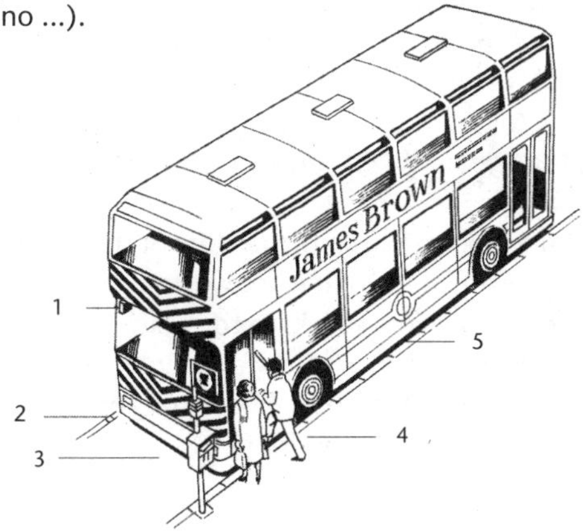

S Situations: What to say when things don't work

James' diary

I'll have to go to Japan in two weeks' time. By the way they fitted my company car out with a global locator system developed in Germany. I took Sheila away for the weekend. We wanted to visit Buxton, a lovely little hill town in the Peak District. Our dream-trip was as follows: Stuck in the bog.

James Kelly (**J**), *Sheila* (**S**)

S: Are you sure we should be relying on this system completely?
J: There's no reason to be alarmed. The people from headquarters insisted that everything was all right.
S: Well, I wouldn't rely on it. It seems to me that it's taking us through some pretty minor roads. *(Sound of crashing through a hedge)* Ouch. What the hell was that?
J: There's nothing to worry about. As you can see the navigator has turned the car right in an attempt to cross this field. Buxton must be on the other side of the field.
S: I wouldn't bet on it. I don't see any town over there. Why are we going slower?
J: I can't say for certain. Probably the built-in petrol saver. It automatically chooses the optimal speed for the most economical consumption of petrol. So there's nothing to be worried about.
S: I see it's a very intelligent system. It's stopped the car altogether. Now we're not using any petrol. What if the Ruver navigator's got its figures all wrong?
J: I haven't the foggiest notion. I'd better get out and have a look. *(splashing through water)* Oh my God! It's driven us into a bog. We're stuck.
S: Aye, stuck in a boggy field in the middle of the Derbyshire countryside and it's getting dark. What are we going to do?
J: I haven't a clue. But don't look on the black side. Every cloud has a silver lining. Why don't we get comfortable and relax?
S: Get comfortable and relax?! Oh, I see what you're driving at. A variation on the "oh, we've run out of petrol theme", is it?

1. Expressing doubt and uncertainty

what if the		it seems to me		are you sure
	a clue		for certain	
notion		bet on		rely on it

Here are some phrases James and Sheila used.
Can you fit in the words from above?

1. .... we should be relying on this system completely?
2. Well, ... that it's taking us through some minor roads.
3. .... Ruver navigator got its figures all wrong?
4. Buxton must be on the other side of the field – I wouldn't .... it.
5. Why are we going slower? – I can't say ....
6. I haven't the foggiest ....
7. What are we going to do? – I haven't ....
8. Well, I wouldn't ....

2. Don't worry!

Sheila is a bit upset and James tries to encourage her.
Do you remember how he does it?

Es gibt keinen Grund zur Aufregung.	There's no ....
Du brauchst dir keine Sorgen zu machen.	There's nothing ....
Sieh nicht immer alles so schwarz.	Don't ....
Alles hat auch eine gute Seite.	Every ....

Time for a smile: Don't worry! Be happy

Doctor: I've received the results of your tests. There's nothing to worry about. I've got good news for you, Mrs McGrady.
Sheila: Doctor – I'm Miss McGrady.
Doctor: Oh, hmmm ... In that case I've got bad news for you.

C Let's s do it crosswise

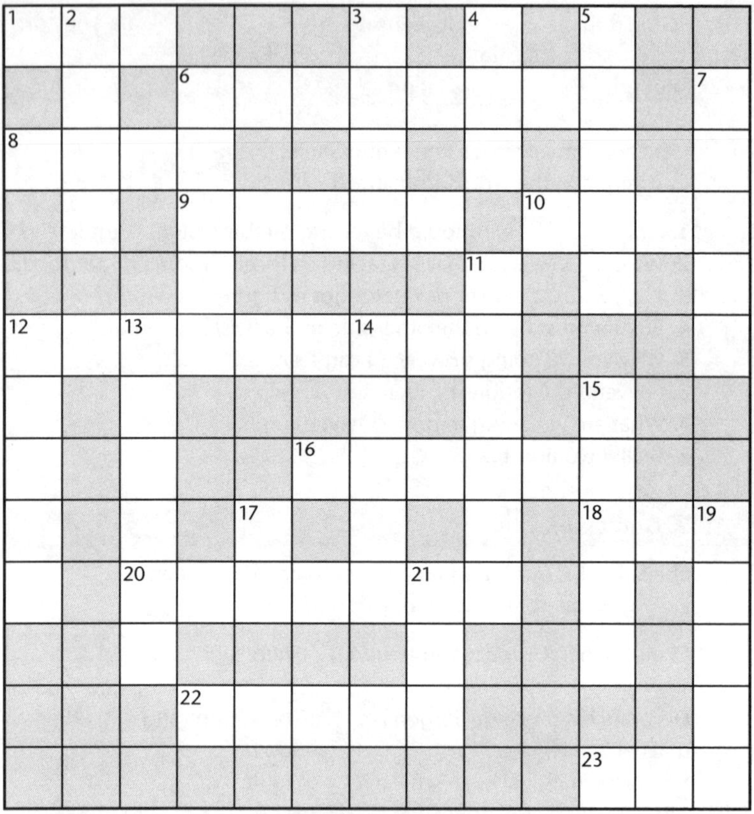

Across

1 complete: state of the – ? –
6 another word for 'plan'
8 another word for 'cover'
9 opposite of 'exit'
10 part of the street (for busses)
12 another word for 'owner'
16 every cloud has a – ? – lining
18 goal, objective
20 another word for 'to control'
22 opposite of 'minor'
23 complete: due – ? – inspection

Down

2 complete: a – ? – of thumb
3 opposite of 'front'
4 public transport
5 another word for 'transmit'
6 short for 'specifications'
7 warn or alarm
11 another word for 'chauffeur'
12 English for 'stolz'
13 design, plan
14 side part of a bus
15 path, lane
17 to arrive – ? – schedule
19 opposite of 'major'
21 Optical Character Recognition

Ⓥ Vocabulary

accuracy	Genauigkeit
alert	warnen
attempt	Versuch
beacon	Lichtstrahl
board a bus	in einen Bus einsteigen
bog	Moos, Sumpf,
bus lane	Busspur
case; a ~ in question	solch ein Fall, ein typischer Fall
clue	Idee, Hinweis
connections	Verbindungen, Beziehungen
consumption	Verbrauch
control point	Kontrollraum
dashboard	Armaturenbrett
deliver	hier: erlösen
depot	Betriebshof
display screen	Bildschirm
driverless bus	Bus ohne Chauffeur
due for inspection	zur Inspektion fällig, zur ~ müssen
Every cloud has a silver lining	Nach Regen kommt auch wieder Sonnenschein
fit a philosophy	zu einer Philosophie passen
get on like a house on fire	blendend auskommen
global locator system	weltweites Ortungssystem
guidance computer	Steuerungscomputer
hedge	Hecke
hitch	Haken; Problem
insist on	bestehen auf
key point	Schlüsselpunkt
locator system	Ortungssystem
magnetic sensor	magnetischer Fühler
magnetic track	Magnetspur
management audit	Überprüfung des Managements
master control panel	zentrale Monitorwand
minor road	Nebenstraße
monitor (v.)	überwachen
navigation system	Navigationssystem
no offence intended	es war nicht böse gemeint, nehmen Sie es nicht übel

notion; I haven't the foggiest ~	ich habe nicht die blasseste Idee
old boys' network	Beziehungen; Vitamin B
Onward!	Weiter (im Text)!
orbit	Erdumlaufbahn
parent company	Muttergesellschaft
pedestrian	Fußgänger
pet; the teacher's ~	Schoßtier; hier: Liebling des Lehrers
petrol saver	Benzinspareinrichtung; Spargang
possessor	Besitzer
predicament	Notlage
proud; the ~ possessor of	der stolze Besitzer von
racialist	rassistisch
rear; at the rear	hinten; im Heck
receiver	Empfänger
relay pictures to the controller	dem Wachhabenden Bilder übermitteln
reliable	zuverlässig
rely on	sich verlassen auf
scan the road	die Straße beobachten
schedule	Fahrplan
arrive on ~	fahrplanmäßig ankommen
scheme	hier: Konstruktion; Plan
select	auswählen
small change	Kleingeld
specs	Spezifikation, Lastenheft
sphere of competence	Verantwortungsbereich
splash through water	durch Wasser platschen
state of the art	Stand der Technik
stuck; be ~	festsitzen
ticket collector	Schaffner
track	Pfad; Spur
tram	Straßenbahn
transceiver	Empfänger
transmitter	Sender
twice; think ~	es sich zweimal überlegen
up to; be ~ ~	vorhaben, im Schilde führen
vibes	hier: Gefühle
worry about	besorgt sein wegen

Chapter 15

What keeps the plane in the air?

- **OUR STORY: Flying the friendly skies**
- **T** **Working with the text:** A little grammar for a change
- **H** **How things work:** Understanding an aircraft
- **S** **Situations:** At the airport
- **W** **Beef up your word power:** Exercises in aviation vocabulary
- **Q** **Our quiz for the expert:** The history of aviation
- **C** **Let's do it crosswise:** An aviation crossword
- **V** **Vocabulary:** Words about flying

Time for a smile

Bad weather forced a plane to keep on circling over the airfield. Finally, the pilot announced: "Ladies and gentlemen, I have some bad news and some good news. The bad news is that we are running out of fuel. The good news is that I am parachuting down to get help."

(*) Flying the friendly skies

James' diary

Holy cow, what a fiasco that was with the navigation system. I guess Sheila thought I did it on purpose somehow or other. From that point of view it's good that I'm going to Japan. Perhaps I can find a system for Ruver. But the other side of the coin is that I'm a bad flier. I can't stand heights. I'm ready to die each time I fly. But I've got to do it. Tomorrow I'm flying from London to Tokyo. Actually I'm getting fed up with being sent around and shouted at. I don't like the present management style. They're more cut-throat than anything we've got back in the States.

James Kelly (**J**), *Jock Wicker* (**W**)

J: *(Sound of airplane engines)* Wow, that was a rough take-off. It felt like a vertical take-off plane. My stomach's still down in my butt somewhere.

W: *(Scottish aero-engineer – very self-confident)* Och, donna worry about that, my boy. It was perfectly normal.

J: Well, I can tell you, sir, that I know a bit about mechanical things – I'm a mechanical engineer, you know – but flying is different. If God had wanted men to fly, they'd have been born with ...

W: *(interrupting)* jet engines in their ...

J: *(interrupting)* Wings. They would've been born with wings.

W: I prefer jet engines. I'm an engineer, too. Aeronautics. My name's Jock, Jock Wicker. Glad to know you.

J: James, James Kelly. Put it there. *(Slapping of hands meeting in greeting).*

W: Right! Well, in a way I can sympathise with you and your fear of flying.

J: Oh, I don't want to make a big thing of it. But, as an engineer, I see how many design mistakes are made in the automobile industry.

W: I know what you mean. I wouldn't be very happy flying in a car, either.

J: Thank god the engineering is more or less perfect for these airplanes.

W: Wrong, my boy. Same percentage of mistakes.

J: Oh, my God. I'm a walking dead man.

W: Sitting actually! Listen. Let me show you a few things. I'm on my way to Tokyo to present a small jet plane we've developed in Glasgow. A completely new design. It can burn all kinds of fuel, even whiskey.

But joking apart. I've got my slides. Do you know what a plane looks like in general?

J: Well, it's just a seat for me.

W: Look at this slide. You can see the different parts and their functions: the fan, the turbines, the compression chamber and the combustion chamber.

J: I don't really understand how such a heavy thing with my fat person could stay up in the air.

W: You know it's basically managed by the jet engine. They're very powerful combustion engines in fact.

J: Well, I know that much.

W: The hot gases they produce are forced out of the back at high speed. That propels the plane through the air.

J: It's not as simple as that, is it?

W: Of course not. These jet engines are also known as gas turbine engines because the hot gases turn a series of blades called turbines. The turbines suck air into the engine and compress the air. Then it is mixed with fuel and burned. Here, look at my transparency. Specially prepared for the Japanese. That type of jet engine is a bit loud. Passenger liners like this have a jet engine with an extra large fan at the front. This muffles the noise.

J: Very interesting. So we're as safe as houses.

W: Well, there are a few things that can happen.

J: Like what?

W: Well, a flock of birds could fly into the fan at the front of the engine. That's enough to bring a plane down. Or there could be something wrong with components supplied. That could cause engine failure. Plane'd fall like a stone.

J: Jesus Christ! My life's dependent on a few feathers and some plastic chips.

W: But I reckon your best chance of getting killed is through human error in traffic control. They mix up flight paths sometimes.

Working with the text

1. Are these statements true (T) or false (F)?

1. An extra large fan muffles the noise. ☐
2. Hot gases are forced out of the back at high speed. ☐
   That propels the plane through the air. ☐
3. Jet engines are not combustion engines. ☐
4. Birds cannot bring down a jet plane. ☐

2. What they might have said but didn't

Replace the these phrases by those used in the text

Let's shake hands.	Put ....
Pleased to meet you.	.... know you.
It's more complex than that.	It's not .... that.
Let's be serious.	But .... apart.
We are absolutely safe.	We're as ....

3. A little grammar for a change

Put the verbs into their correct form.

1. If God (*want*) men to fly, they (*be born*) with wings.
2. I (*not be*) very happy flying in a car, either.
3. There (*can*) be something wrong with components (*supply*).

Time for a smile

The holiday charter flight had its usual number of children. One little boy in particular was getting on everybody's nerves by running up and down and upsetting drinks. An elderly gentleman grabbed him by the arm and said: "Sonny, do me a favour. Please go and play outside."

H How things work in the plane

A box of jet verbs

suck in	einsaugen	flow	strömen
draw in	einsaugen	cool	kühlen
propel	antreiben	heat	erhitzen
contain	enthalten	mix	vermischen
enter	eindringen	drive	antreiben
increase	erhöhen	speed	rasen
rotate	sich drehen	pass	passieren
expand	sich ausdehnen	eject	ausstoßen
push	schieben	burn	verbrennen

1. Verbs in action: The jet engine

Try to use all the verbs. The first has been done for you.

The engine that propels big airliners .... air in and .... it at high speed. At the front of the engine, at the intake, a large fan .... to .... air in. Some of the air .... the engine. The mainstream .... the compressors, which .... both rotating and stationary blades. The compressors .... the pressure of the air, which then .... to the combustion chamber where it is .... with kerosene or paraffin. The mixture .... explosively and .... The hot, high-pressure air .... towards the exhaust with tremendous force and .... the plane through the air. On its way the hot gas .... through turbines that .... the fan and the compressors.

Time for a smile

The new Super Airbus 2000 on its maiden flight to New York had taken off when a soft voice came over the speaker system: "Welcome on board our Super Airbus 2000. This is a fully computerised aircraft, without pilot, co-pilot, stewards and stewardesses. No need to worry, everything is under control. There is nothing that can go wrong ... go wrong ... go wrong ... go wr ... go ... go."

Toolbox : The anatomy of a Being 747

1	radar equipment	Radar
2	flight deck	Cockpit
3	wing	Tragflächen
4	aileron	Landeklappe
5	undercarriage	Fahrwerk
6	jet engine	Düsentriebwerk
7	fuel tank	Treibstofftank
8	tailplane	Heckflügel
9	elevator	Höhenruder
10	tail fin	Heckflosse
11	rudder	Seitenruder

And now put the right letter into the circles

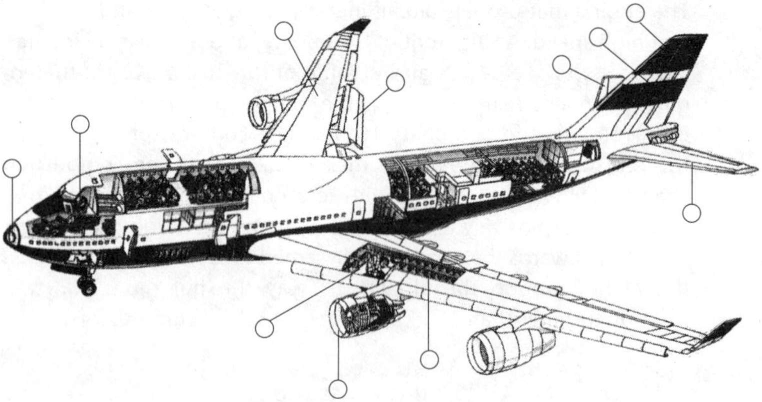

Time for a smile

One charter flight was so crowded – the pilot went by boat.

S Situations: At the airport

Choose the best alternative.

1. Excuse me, can you tell where the .... is, please?
A check-in desk **C** booking office
B reception desk **D** ticket counter

2. Good morning, I'd like to .... from New York to London, please.
A book an early flight **C** sign in on an early flight
B book an early passage **D** apply for an early flight

3. Which .... from?
A hall do I start **C** lounge do I go on board
B gate do I depart **D** platform do I board

4. Excuse me, where is .... ?
A the departure hall **C** the waiting room
B the boarding lounge **D** the departure lounge

5. What are my .... ?
A customs-free allowances **C** duty-free shoppings
B duty-free allowances **D** tariff-free allowances

6. How many .... do you have?
A items of luggage **C** peaces of luggage
B pieces of suitcases **D** lumps of luggage

7. Don't .... your hand luggage .... at any time.
A leave ... unattended **C** let ... unobserved
B leave ... unwatched **D** let ... unguarded

8. Would all passengers for flight D 492 please .... gate 52 immediately.
A be going **C** proceed to
B move to **D** head for

9. Flight number D 492 to London is now .... gate 25.
A boarding at **C** boarding in
B starting from **D** taking off

10. .... Passengers for London on flight D 492, please go to exit.
A Last warning. **C** Final upcall.
B Final calling. **D** Last call.

Beef up your word power

1. One verb is different. Mark it!

enter	propel	contain	flow
eject	push	mix	speed
suck in	drive	burn	heat
take in	rotate	heat	pass

2. Three parts are in the same area; one is not. Mark it!

fan	rudder	tail fin
turbine	fuel tanks	elevator
compressor	engine	radar
undercarriage	aileron	rudder

3. Where do these components belong?

	aileron		fan
....	tanks	....	compressor
	engine		turbine

4. What is 'it' used for?

It causes the airliner to fly a curve. ....

It is needed during take-off and touch-down. ....

It is the part reserved for the pilots. ....

It can enlarge the surface of the wings. ....

It causes the aircraft to gain or lose height. ....

It has a curved surface and produces the lift as it moves through the air. ....

Time for a smile

Air traffic controller: What is your height and position?
New pilot: I'm about five feet ten inches and I'm sitting in the pilot's seat.

Our quiz for the experts

1. Who were the first people to fly?
   A Chinese criminals
   B Charles Lindbergh
   C the Montgolfier brothers
   D Leonardo da Vinci

2. Where and when was the first airship built that could carry passengers?
   A by the Montgolfier brothers 1783 in France
   B by the Baron von Münchhausen in 1845
   C by Aristotle in Egypt 250 BC

3. Is it true that the Frenchman Jacques Charles invented the first balloon that was steam-powered?

4. The curved surface of the wing (aerofil)produces the lift that keeps the plane in the air. This principle was discovered by
   A Leonardo da Vinci
   B the tailor of Ulm
   C Daniel Bernouilli
   D the Chinese Lao Min

5. The first use of the principle of the wing (aerofil) was in
   A gliders
   B ships
   C kites
   D cars

6. Who flew the first modern airplane powered by a petrol engine?
   A von Richthofen
   B the Wright brothers
   C Charles Lindbergh
   D Paul Cornu

7. When and where was the first helicopter invented?
   Who was the inventor?

Time for a smile

The captain of the Boeing 747 had forgotten to switch off the speaker system so the passengers could overhear the captain in the cockpit: "I'll have a cup of coffee and then make love to that new stewardess." The young stewardess at the other end of the plane turned red and ran down the cabin to warn the crew that the speaker system was on. But in her haste she tripped and fell. An elderly lady helped her get up. "There's no need to hurry", she said. "Let him finish his coffee first."

C Let's do it crosswise

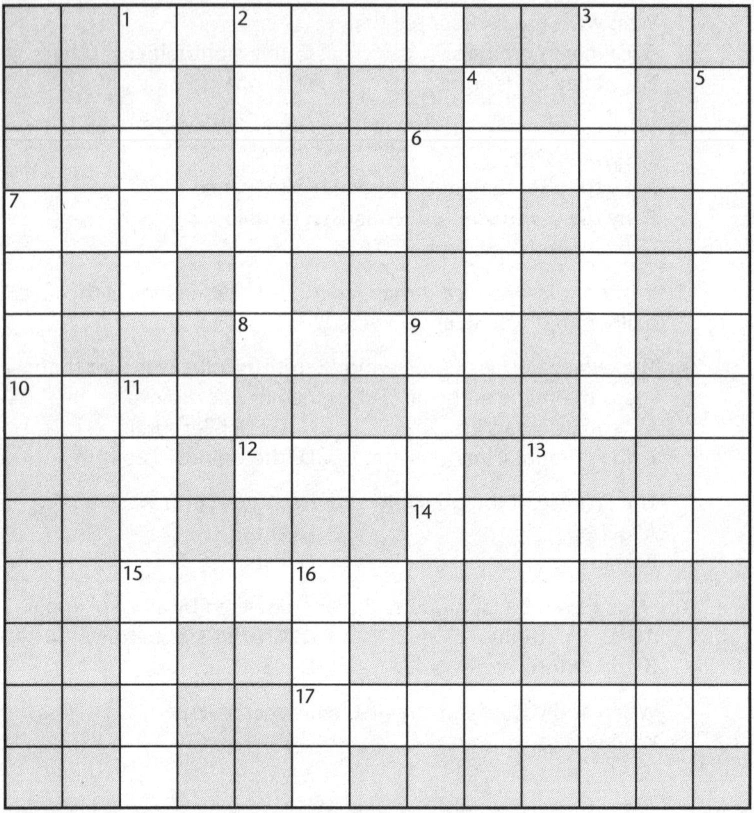

Across

1 – ? – LH452 is now boarding
6 birds have it, too
7 he serves you whisky and beer
8 English for 'ausstoßen'
10 flying machine used in China
12 airplanes wait there for you
14 aviation pioneer
15 part of a turbine
17 another word for aircraft (USA)

Down

1 James is a bad – ? –
2 where the air is drawn in
3 another word for luggage
4 he flies an airplane
5 another word for airplane
7 it means 'to draw' in air
9 team on a plane
11 blades are mounted on to it
13 both planes and fish have it
16 it means to 'suck' in air

Ⓥ Vocabulary

aeronautics	Luftfahrt
air traffic controller	Fluglotse
airfield	Flugfeld
as safe as houses	absolut sicher, bombensicher
aviation	Luftfahrt
blade	Flügel-; Schaufelblatt
butt	Hintern
coin ; the other side of the ~	die Kehrseite der Medaille
compress the air	die Luft komprimieren
cut-throat; be ~ ~	Halsabschneider sein
dependent on	abhängig von
exhaust	Auspuff
explorer	Forscher
fan	Ventilator, Lüfter
fed up; be ~ ~ with	die Nase voll haben von
flier	Flieger (Flugpassagier)
flight path	Flugschneise
flock of birds	Schar Vögel
glider	Segelflugzeug
height	Flughöhe; Größe
intake	Ansaugöffnung
inventor	Erfinder
kerosene	Kerosin (Flugzeugkraftstoff)
kite	Drachen
maiden flight	Jungfernflug
muffle the noise	das Geräusch dämpfen
overhear	mithören
parachute	mit dem Fallschirm abspringen
particular	besonders
percentage	Prozentsatz
point of view	Gesichts-, Standpunkt
proceed to	sich begeben nach
propel	antreiben
punishment	Strafe

purpose; on ~	absichtlich
put it there	schlagen Sie ein; hier meine Hand
reckon	annehmen, vermuten
reliable	zuverlässig
self-confident	selbstbewußt
slides	Dias; hier Folien
stand; I can't ~ heights	vertragen; Ich kann keine Höhen ~
stationary	starr, fest, unbeweglich
steam-powered	dampfgetrieben
surface	Oberfläche
sympathise with	mitfühlen mit
take off	starten; abheben
take-off	Start
touch-down	Landung
trip	stolpern
unattended	unbewacht
vertical take-off plane	Senkrechtstarter
whirl	wirbeln
wing	Flügel; Tragfläche

Jobs in the World of Technology

Chapter 16

Congratulations! You've made it to the last chapter, which will be different from all the other chapters.

(✻) **OUR STORY: Making money on the Internet?**

(R) **Extra reading:** What James should know about the Internet

(V) **Vocabulary:** Words about jobs

Food for thought

In a joint press conference in St. Peter's Square, Microsoft Corporation and the Vatican announced that the computer software company will make religion interactive, easier and more fun for a broader range of people. "We expect a lot of growth in the religious market in the next five to ten years", said Gates. "The sacraments will be available on-line for the first time. You can get communion, confess your sins, buy inexpensive indulgences and receive absolution and thus reduce your time in purgatory – all without leaving your home."

(*) Making money on the Internet?

James' diary

I liked it better in the old days of Ruver and Gonda. The atmosphere has changed and somehow the old values are not respected. Everything AMW is concerned with is performance. I especially don't appreciate the intrigues and the attitudes of some of the "little" bosses. I talked to Sheila about the idea of resigning from the company.

In Sheila's apartment; James (**J**); *Sheila* (**S**)

J: *(fade-in)* ... so basically I'm on the look-out for a new position. Something in engineering or something technical. I've been looking at the ads in the dailies.

S: Well, did you find anything?

J: There was one thing as an advisor in Russia. Miles away.

S: Oh, I wouldn't much fancy that. After all, we've only just got to know each other.

J: There were other things such as toolmakers, maintenance fitters, welders etc.

S: That's not your level. You're a graduate. You couldn't take anything less than an engineering position.

J: Anyway there was nothing for a design engineer. That would have been most interesting for me. Creating new products and developing existing ones. That's engineering. Finding new applications for things. That's where it's at. And I've got some experience in matters of quality and reliability. Of course there is production engineering, too, but that has more to do with processes and handling. Finding flaws and defects and correcting them. A production engineer gets ideas from the design departments and has to decide if it is cost-effective and practicable. That's not what I want.

S: Look, Jamie, I know it's not my business but, well, since you've been so depressed at being pushed around at this company, I've been looking round, too. And I found this ad on the Internet. I've been practising with my computer a bit recently. Like you told me to. "Use it as a tool" you said. Take a look at it.

J: Oh great!, You do surprise me, Sheila. You're really enterprising.

Quality Engineer c.£ 30 K

Our manufacturing facility at Long Eaton near Nottingham commenced production in early 1946. The workforce has expanded rapidly and we now require a quality engineer to join us. Reporting to the Director of Quality Assurance, you will be responsible for ensuring quality-related activities.
You are a mechanical engineering graduate with experience in the automotive industry. You should have a working knowledge of CAD, CAPP and problem-solving techniques.
A deep knowledge of international quality systems, ISO 9001, GS, CE etc.
Plus-points: attractive salary, benefits. Excellent opportunities for career development.
To apply, please send a CV stating current salary to: ...

S: Look, there are some points that fit you quite well.
J: Quality? That's not really my thing.
S: Just keep in mind it's appropriate to your level. Don't go backwards! Take more responsibility! What about advantages? First of all it is in the automotive industry. That's your thing. Next, it's a leading engineer. That's what you want. You love trouble-shooting and problem-solving! What have you been doing at Ruver if it's not problem-solving? And I don't want to put you under pressure, but it's here in England. That means we could stay together.
J: My god! That was quite a speech. I know what you mean, but in principle we shouldn't let relationships interfere with business. We're all free agents, aren't we?
S: Of course we are, James. I was just listing the plus-points.
J: Okay, it looks good, but I have to say I've had enough of those rude bosses with tunnel vision. I want to be myself. And they don't respect that.
S: Well, you can't say I haven't tried. You don't seem to want to stay here and I must say it annoys me.
J: I need to do my own thing. I can't help it. I need a new challenge.
S: Well, if you had any courage, you'd resign and build up your own business.

J: That's easier said than done. You need funds for that. Money. Advertising. You name it.
S: That's just an excuse. There are enough people founding companies and doing things on the Internet. I ... *(James interrupting)*
J: Oh, that's it, Sheila! You've hit the nail on the head. The Internet. We're going to build a business on the virtual waves. Then we don't need so much capital. I see us now, making money on the Internet! That's where the future is!
S: Lord save us! What have I started now?

What James should know about the Internet

What is the Internet?

The Internet is considered as a way for the greatest minds in academia to be in contact with each other. It is not true that the Internet was invented by Bill Gates. Although he will end up having the exclusive rights, just as he already owns the exclusive rights for the library of the Vatican and the major art galleries.

Who is 666?

Bill Gates signs all his memos with 666. This has nothing to do with sex. The computer expert will have realised that spelled out in binary code his name is the most harmonious name in the beautiful world of electronics. Why? Bill's real name is William Henry Gates III. By converting the letters of his name into ASCII code and adding the III, you get the following:

B	66	G	71
I	73	A	65
L	76	T	84
L	76	E	69
plus III	3	S	83
Total	**294**	**Total**	**372**

294 + 372 = 666

A coincidence? Just wait and see. Bill Gates invented WINDOWS95. Once again, convert this into ASCII and add the figures up. And what do you get?

87 + 73 + 78 + 68 + 79 + 87 + 83 + 57 + 53 + 1 = **666**

What do you need in order to use the Internet?

The Internet is a vast gov/netw.ork led by mice. If you keep pressing your mouse you can keep travelling in cyberspace until Telecom disconnects you because your telephone bills exceed your salary. It is a communication miracle that allows you to communicate with everyone on html// http// planet and get access to information you never dreamed you would need.

A modem connects your computer to the Net. It will move you very, very slowly through cyberspace. But the trick here is that when you are ever so slowly downloading naked pictures of Julie Foster, Julia Roberts or Arnold Schwarzenegger, you can also iron your shirts.

What is a home page?

When your mouse leads you onto the Web, you will visit the mostly useless sites with a home page. Anybody can have a site, and thus a home page that pollutes cyber space. Once you reach a home page you can click on the various links to find the treasure chest of goodies waiting for you. Now, let's have a look at a typical homepage:

Welcome to my personal home page

About me:

I am an inmate of Sing Sing Prison.
As a result of an error of justice I was recently condemned to the electrical chair for rape on a large scale and murder of three drunk tourists from Holland who peed against my car.

❖ click on here to read all the newspaper reports about this judiciary scandal.

❖ click on here if you want transfer some money to my bank account. I need it to appeal against my sentence to the Supreme Court.

My hobbies:

My hobbies are knitting, gardening and writing my memoirs.

❖ click on here to learn more about my hobbies.

My favourite books:

The Truth about Jack the Ripper · Frankenstein · Jerry Cotton

My e-mail address: ripboy@sing.bighouse.edu

What exactly is E-mail?

E-mail is on a par with Gutenberg's invention of the printing press. It's the future of mankind. It will eliminate the Postal service, fax machines, telephones and bicycle messengers. E-mail will raise the culture of letter writing to a higher intellectual level

Here is a look at a typical e-mail message.

> From Hannelore@priv.apartmntkanzleramt.gov
> Wed Apr 19 12:23:46 1998
> Return-Path: <Hannelore@priv.apartmntkanzleramt.gov
> To: hKohl@priv.apartmnt.kanzleramt.gov
> Subject: Anti-stress training, slimming
>
> I'll be playing badminton with Seehofer's toupee at 4:32 on the lawn behind the Amt. Little Blühm will be there, too.
> Are you free?
> *Hannelore*

How to read an e-mail on the World Wide Web

http	hight-tech trash pile
xx	kisses
<ns>	Not smiling
<ckcph>	Can't keep chatting, prostrate hurts
GOOMW	Get out of my way
<tush>	Totally useless stuff here

Why you should think twice before buying a computer

1. Any given programme, when running, is out of date.
2. The up-date does not run properly on your old computer.
3. As soon as you have got used to the software, they offer you an update that takes months to get used to.

Ⓥ Vocabulary

ad	Anzeige
advisor	Berater
annoy	ärgern
apply	anwenden; sich bewerben
appreciate	mögen, schätzen
appropriate	angemessen
attitude	Haltung, Einstellung
available	verfügbar
broader range of people	breiteres Publikum
CAD	Computer Aided Design
CAPP	Computer Aided Process-Planning
CE	Certificat Européen
coincidence	Zufall
commence	beginnen
concerned; be ~ with	sich kümmern um
consider as	halten für
dailies	Tageszeitungen
defects	Fehler, Defekt
design engineer	Entwicklungsingenieur
ensure	sicherstellen, gewährleisten
error of justice	Justizirrtum
exceed a salary	ein Gehalt übersteigen
expand rapidly	schnell wachsen
fancy, I don't ~ that	das gefällt mir nicht
fit someone well	gut auf jemanden zutreffen
flaws	Fehler
free agents	freie Menschen
funds	Geldmittel
goodies	Bonbons, Orchideen
graduate	etwa: Hochschulabsolvent
GS	geprüfte Sicherheit
handling	Handhabung
indulgence	hier: Ablaß
inmate	Insasse

interfere; relationships shouldn't ~ with business	Privates und Geschäftliches sollte man nicht vermischen
iron (v.)	bügeln
ISO	International Standards Organisation
joint press conference	gemeinsame Pressekonferenz
judiciary scandal	Justizskandal
K; £ 30 K	30,000 £ (K = Kilo)
knitting	Stricken
level	hier: geistiges Niveau
look round	sich umschauen
Lord save us!	Gott sei uns gnädig!
mice (plural of mouse)	Mäuse
miracle	Wunder
par ; on a ~ with	auf gleicher Ebene wie
pee	pinkeln
problem-solving	Problemlösen
prostrate	Prostata
purgatory	Fegefeuer
rape	Vergewaltigung
receive absolution	Absolution erhalten
related; ~ activities	verwandte Tätigkeiten
reliability	Zuverlässigkeit
report	berichten; unterstehen
require	benötigen; fordern
resign from a company	aus einer Firma ausscheiden
scale; on a large ~	in großem Maßstab
site	Standort
speech	Rede; Ansprache
spell	buchstabieren
state current salary	das augenblickliche Gehalt angeben
toolmakers	Werkzeugmacher
treasure chest	Schatzkiste
tunnel vision	Scheuklappen; beschränkter Horizont
used; get ~ to	sich gewöhnen an
values	Werte
virtual waves	hier: in der virtuellen Welt

Key to Exercises

CHAPTER 2

T Working with the text

1. Are the following statements true (T) or false (F) ?

1.	James is in England for the first time.	**F**
2.	The company taxi has an automatic gear-box.	**F**
3.	There's something wrong with the electronic system.	**T**
4.	The taxi is a test car.	**T**
5.	The defects are repaired at the garage.	**F**

2. What they might have said but didn't

1.	The engineers can't read the *instructions.*	specifications
2.	We have to *stop* at one of our garages.	drop in
3.	I'll just *turn on* the windscreen wipers.	switch on
4.	Defective *parts* are marked with an X.	components
5.	The car's still full of *preliminary* problems.	teething
6.	Ruver cars are *the best.*	tops

H How things work in the car

1. Tools in use: What to do if you want to ...

show a new direction	use the indicator
warn a pedestrian	sound the horn
avoid an obstacle	turn the steering wheel
clean the windshield /- screen	switch on the windscreen wipers
drive at night	switch on the headlights
see what's going on behind	look in the rear-view mirror

2. Tools in use: Complete!

you want to stop the car	put the brakes on
you want to go faster	step on the accelerator.
you want to go into neutral	push in the clutch.

Situations

1. In case of breakdowns - don't worry, be happy

Typical problems ...	and what say at the garage
1 you've run out of petrol	K fill her / it up
2 there isn't enough oil	I top / fill it up
3 the clutch is slipping	E adjust it
4 you have a puncture	A put on / fit the spare
5 the battery is flat	C recharge it / charge it up
6 the radiator is leaking	B make it watertight / seal it
7 the car won't start	D try to bump-start it
8 the fanbelt is slipping	F tighten it
9 the motor is too loud	G weld the hole in exhaust pipe (or the silencer)
10 the oil is leaking	H try to seal the sump
11 the engine sounds rough	J adjust the ignition / check the plugs

Be polite and add 'please' each time you ask for help.

2. Wrong order: a driving test

4 Push in the clutch.
7 Press the accelerator, drive away.
3 Release the handbrake.
5 Put the car into first gear.
8 Push the clutch in again.
6 Release the clutch very slowly.
2 Turn it and start the engine.
6 Let the clutch out.
1 Put the key into the ignition.
9 Shift / change gear and so on.

Well done! However you're driving on the wrong side of the road. We're in Britain.

3. Do it yourself – Repairing the puncture

2 Open the boot.
7 Jack up the car.
4 Detach the hubcap with a screwdriver.
3 Open the tool kit and take out the wheelbrace and the jack.
1 Apply the handbrake.
6 Loosen the wheelnuts, but don't unscrew them completely.
9 Put on the spare.
12 Take the wheelbrace and tighten the screws
14 Stop at the next petrol station and check the air pressure.
5 Take out the spare wheel.
13 Replace the hubcap.
8 Unscrew the wheelnuts completely; remove the wheel.
10 Replace the wheelnuts and hand-tighten them.
11 Lower the car.

There you are!

W Beef up your word-power

1. Odd man out

A	B	C	D
brake	tyre	radiator	antenna
pedal	hubcap	carburettor	glove compartment
clutch	wheelnuts	boot	dash board
accelerator	jack	battery	gear shift

2. In other words

Find a word that has (almost) the same meaning

component	part
switch on the headlights	turn on the headlights
let out the clutch	release the clutch
take off the hubcap	detach the hubcap
fill up the oil	top up the oil
change gear	shift gear

3. Explain it with the opposite

lower the car	jack up the car
switch on the headlights	switch off the headlights
slow down the car	accelerate
release the clutch	push in the clutch
screw it on	unscrew it
discharge the battery	recharge the battery
loosen the wheelnuts	tighten the wheelnuts

Q Our quiz for the experts

Do you know the combustion engine?	Yes	No
It converts heat into motive power.	Yes	
Jets and rockets have it, too.	Yes	
A piston rotates inside a cylinder.		No
Strong vehicles have two-stroke engines.		No
All combustion engines have spark plugs.		No
Diesels have inlet and outlet valves.	Yes	
Four-stroke engines have four valves.	Yes	
Combustion engines burn air and petrol.	Yes	
74 kWh is about 100 horsepower.	Yes	
Cars waste 60% of their energy.	Yes	

C Let's do it crosswise

	[1]S	[2]P	A	R	[3]E		[4]B	[5]O	O	T		
		E			N			U			[6]J	
[7]A	U	T	O		[8]G	R	[9]A	T	E		[10]A	D
U		R			I		C				C	
[11]T	O	O	L		N		[12]C	H	E	[13]C	K	
O		L			E		E			H		
M							L			A		
[14]O	F		[15]B	R	O	K	E	N		N		
B			R				R			G		
I			A			[16]B	A	T	T	E	[17]R	Y
[18]L	E	A	K				T				U	
E			E			[19]W	E	L	D		N	

CHAPTER 3

T Working with the text

1. Are these statements true (T) or false (F)?

1.	Today James is in the R & D Department.	F
2.	The director sent him there to help to get things off the ground.	F
3.	Mr Bennet thinks that American cars are uneconomical.	T
4.	Mr Bennet thinks that the new four-door range is fine.	T
5.	The new four-door range is a sports car.	F
6.	James wants to integrate the horn in the rim of the steering wheel.	T
7.	James thinks that speedometers with hands are old-fashioned.	T

2. What they might have said but didn't

1. He *thought* he knew all about the defects.	He *reckoned* he knew all about the defects.
2. The director sent me down to *gather* information.	The director sent me down to *collect* information.
3. Perhaps the American *method* can help to get things *started.*	Perhaps the American *approach* can help to get things *off the ground.*
4. Your problem is *obvious.*	Your problem is *staring you* in the face.
5. You don't *understand* me.	You're not *getting* my *message.*
6. *Wait* a minute!	*Hold on* a minute!
7. I don't agree.	I can't *go along* with this.
8. We'll *solve this problem.*	We'll *beat this trouble.*

H How things work in the car

1. Tools in use: Imagine you are a mechanic

1.	**B.**	The speedometer is	in the instrument-panel
2.	**A.**	The gear-lever is	right of the driver's seat
3.	**D.**	The dials are	in the speedometer
4.	**C.**	The instrument-panel is	in the dashboard

2. Tools in use: What is the 'it'?

It shows how much petrol you've left.	the petrol gauge
It is polished regularly in Germany.	the bodywork / paintwork
It consists of several meters..	instrument panel
It has a glove compartment.	dashboard
It makes the seats more comfortable.	upholstery
It is a meter that shows your speed.	speedometer / speedo
It is a meter that shows your mileage.	milometer

S Situations

1. At a petrol station

Match phrases and	**parts of the car**
1. A. Add some distilled water.	battery
2. C. Could you wax it for me?	paintwork / bodywork
3. D. The bulb's gone.	headlight
4. E. Could you give me some anti-freeze?	radiator
5. B. They are worn. They need changing	tyres

2. In the Design Department

The old design	**The new design**
is too big	must be smaller
consumes too much petrol	must be more economical
is too heavy	must be lighter
is a bit old-fashioned	must be more modern
looks like a family car	must look sporty

W Beef up your word power

1. What do you get when you put these together?

	nuts		plugs		water
wheel	tyre	engine	piston	battery	anti-freeze
	hubcap		valve		cells
	polish		jack		bonnet
paintwork	wax	tool kit	brace	bodywork	boot
	wash		screwdriver		wing

2. Intelligent guessing

1. The *bulb* in the headlight has gone.	Birne
2. The tyres are *worn*. They need changing.	abgenützt
3. There's *a dent* in the door. Can you fix it up?	Beule
4. Could you give me some *anti-freeze*?	Frostschutzmittel
5. Can you do a 10,000 mile *maintenance check*?	Inspektion
6. The *petrol gauge* is defective.	Benzinuhr
7. Do you have a *car wash*?	Waschanlage

Q Our quiz for the experts

1. Carl Benz invented a combustion engine in 1880.
2. Rolls Royce, Aston Martin; Jaguar; Rover.
3. **A.** After the First World War.
4. Ford invented the assembly line.
5. **C.** catalytic converters.
6. Detroit and Cleveland.
7. Airbag and seat belt.
8. **C.** Petroleum is used in lamps.

C Let's do it crosswise

(1) R	I	(2) M		(3) S	P	E	E	(4) D		(5) G	A	S
A		I						E		A		
D		(6) L	O	(7) W		(8) C	E	N	T	U	R	Y
I		O		A		I		T		G		
A		(9) M	E	T	E	R			(10) K	E	Y	
T		E		E		C			E			
O		T		R		U			R			
R		E				L		(11) F	O	R	D	
		R		(12) F	L	A	T		S			
				U		R			E			
	(13) T	Y	R	E			(14) P	A	N	E	L	
				L					E			

CHAPTER 4

Working with the text

What they might have said but didn't

Replace the words in italics by those used in the text

1. It was the meeting to *solve* the *problem* of the engine.	It was the meeting to *hammer out* the *matter* of the engine.
2. I *suggest* we look at our carburettor engine again.	I *propose* we look at our carburettor engine again.
3. *Let's* get down to business.	*We'd better* get down to business.
4. Could you *project* it on the OHP?	Could you *flash* it *up* on the OHP?
5. If the air's not clean, that'll *create* other problems.	If the air's not clean, that'll *cause* other problems.

Tools in use – Inside the engine.

It sends sparks to plugs.	distributor
It is long , oily and made of metal.	dip stick
It is a small power station.	dynamo
It transforms fuel into gas.	carburettor
It is only in diesels.	glower plug
It is a car for safaris.	off-road car

Situations

1. Making suggestions at a meeting

1. May I *submit* a proposal?
2. My *advice* is to make the car look more sporty.
3. I strongly *recommend* that we use a boxer motor.
4. *What about* changing the design?
5. Don't you think fuel injection *might be* a good idea?
6. *Wouldn't* it be wiser not to use old materials?
7. I *advise* you to throw out all old components.
8. I would *suggest* that we put off the decision.

2. Complaining to the sales department

Causes	and their effects
Mick was driving off-road.	The exhaust pipe fell off.
The distributor is badly out of time.	The motor doesn't run smooth.
There are holes in the exhaust.	The motor backfires.
The carburettor needs setting.	The engine drinks too much.
The gasket is worn.	The oil sump is leaking.

3. Technical defects require more than fixing

Learning by doing. Fit in the missing links:

Complaint	and Apology
1 I've a *complaint* to make.	I'm *awfully sorry* to hear that.
2 I really want *to complain* most strongly.	Please accept our *apologies*,
3 I'm utterly *dissatisfied* with the performance of my Ruver.	I can understand why you are *angry*.
4 I really can't *accept* this kind of thing again.	Let me *apologise* for the situation. I'll look into it immediately.

Beef up your word power

1. Odd man out: What does not belong into the list?

A	B	C	D
breakdowns	replace	loose	anti-freeze
worn components	check	tight	exhaust fumes
over consumption	change	worn	cooling liquid
broken parts	repair	noisy	gasoline

2. Intelligent guessing

Which parts of the diesel engine belong together?

1	piston	**B**	ring
2	radiator	**D**	cooling-liquid
3	dynamo	**E**	cable
4	compression chamber	**A**	glower plugs
5	carburettor	**C**	atomiser

Our quiz for the experts

1. Which device avoids pollution of the air? **(A)** It's the catalytic converter.
2. In a Wankel engine the pistons rotate instead of moving up and down. It's true.
3. What is the shape of a Wankel piston? It's **(B)** triangular.
4. The car's worst enemy is **(B)** friction, of course.
5. What is the effect of oil in an engine? **(A)** lubrication (Schmierung).
6. If you press the accelerator you open **(C)** the throttle.
7. What is injected into the cylinder? **(B)** vapour of air and fuel.

Let's do it crosswise

		1 M		2 S	P	E	E	3 D		4 G	A	S
5 F	A	I	L					E		A		
U		L		6 P		7 C	E	N	T	U	R	Y
E		O		E		I		T		G		
L		8 M	E	T	E	R			9 K	E	Y	
		E		R		C			E			
		T		O		U			R			
		E		L		L		10 F	O	R	11 D	
12 T	Y	R	E		13 F	A	N		S		E	
U				14 O		R			E		V	
R				P					N		I	
N			15 R	E	V	E	R	S	E		C	
				L							E	

CHAPTER 5

Working with the text

1. Are these statements true (T) or false (F)?

1. Ruver produces most of the components themselves.	F
2. Tools are adapted to new designs in the erection shop.	T
3. Mud-guards are formed on shaping machines.	T
4. Coarse threads are manufactured on the lathe.	F
5. Welders use gas torches.	T

2. What they might have said but didn't

1. I'd like to give you an *idea* of what we're using.	I'd like to give you an *overview* of what we're using.
2. Where did they *find* all this ancient equipment?	Where did they *dig up* all this ancient equipment?
3. *Excuse me, but can you see* those gas bottles over there?	May I *draw your attention* to those gas bottles over there?
4. We are proud to *have* such sophisticated equipment.	We are proud to *possess* such sophisticated equipment.
5. It won't *function* without the gas.	It won't *work* without the gas.
6. If we can *return* to the more important matters.	If we *can get back* to the more important matters.

How things work in the factory

1. Tools in use: Match numbers and letters

1 drilling machine	D You make small holes with it.
2 mould	C You pour liquid metal in it.
3 shaping machine	B It gives sheet iron its final form.
4 bracket crane	E You lift heavy objects with it.
5 lathe	A It shapes a workpiece by turning it.

2. Shops that don't sell anything

These are our suggestions.

1.	The one time in the day you lean back and relax is the one time the Boss walks in?	**A**
2.	If it were be cheaper to buy a new unit, the company will insist on repairing the old one.	**D**
3.	The motor will rotate in the wrong direction.	**C**
4.	When an error has been detected and corrected, it will be found to have been correct in the first place.	**C**
5.	After the machine tool has been assembled, extra components will be found on the bench.	**B**
6.	If a machine tool finally functions perfectly all the other production units will malfunction.	**B**
7.	Any wheelnut, when dropped will roll into the least accessible corner.	**D**
8.	After the last of 16 screws has been tightened, it will be discovered that the gasket has been forgotten.	**D**

3. Tools in use: What is the 'it'?

It is used to give a sheet of iron a form.	shaping machine
It is a machine to make pistons with.	milling machine
It joins parts with an alloy of tin and lead.	soldering iron
It uses electricity to join larger parts.	welding rod

S Situations: How to interrupt

May I *interrupt* you for a moment?
Pardon me for *interrupting*, but is this really relevant?
Excuse me, could I just *break in* here ...?
If I could just *continue*, I'll answer your question in a minute.
Coming back to what I was trying to explain ...
To *go back* to what I was saying ...
To *return to* my subject ...
If I might just *finish* my sentence.

Beef up your word power

1. What do you get when you assemble the components?

	handle		grip		handle
hammer	head	saw	blade	knife	blade
	claw		teeth		sharp edge

2. The Ostfriesen final exam for engineers

1	What do you dig holes with in the garden?	**E** spade
2	What do you chop wood with?	**D** axe
3	What do you measure your height with?	**C** rule
4	What do you drive in nails with?	**A** hammer
5	What do you need to check surface level?	**B** level

Our quiz for the experts

1 gloves **2** pliers **3** spanner **4** drill **5** trowel **6** screwdriver **7** jig saw **8** hammer

Let's do it crosswise

1 L	A	2 T	H	3 E		4 F	A	U	5 C	E	6 T	
E		I		R					H		7 O	N
V		D		R				8 J	O	I	N	
E		Y		O					P		G	
L			9 G	R	I	10 P		11 C		12 C	U	T
			A			I		H			E	
13 H	O	I	S	T		14 P	L	I	E	15 R	S	
A			K			E		S		U		
N		16 L	E	A	K			E		L		
D			T			17 W	E	L	18 D	E	R	
L				19 S					I			
E			20 C	A	S	T	I	N	G			
				W								

CHAPTER 6

T Working with the text

1. Are these statements true (T) or false (F)?

1.	Ruver and Gonda have negotiated a joint venture.	T
2.	Ruver engines will be replaced by ones from Gonda.	T
3.	James is sent to Indonesia to buy parts for Gonda engines.	F
4.	The Indonesian reactor uses nuclear fusion.	F
5.	Mr Maylay expects James to talk to the government.	T
6.	He thinks of building their own nuclear power plant.	F

2. What they might have said but didn't

1.	Mr Kelly, you *have* to attend a meeting at 10.	Mr Kelly, you *are* to attend a meeting at 10 a.m.
2.	All engines *are* to be replaced by ones from Gonda.	All engines *will* be replaced by ones from Gonda.
3.	Thank-you for your *help*.	Thank-you for your *support*.
4.	That's *the way* the power is produced in Indonesia.	That's *how* the power is produced in Indonesia.
5.	We *have to look* for other sources of energy.	We *must search* for other sources of energy.
6.	That's *the reason for your presence.*	That's *why you are here.*

H How things work in the power plant

1. Can you match these parts with the things in our story?

1	wheel	**C** turbine
2	wire ring	**A** coil
3	coil and magnet	**D** generator
4	waterfall	**B** chute

2. How the electricity travels to your home

1 power plant, **2** power line, **3** pylons, **4** transformers, **5** voltage, **6** power lines, **7** transformer, **8** transformer, **9** alternating current, **10** coils, **11** magnetic field, **12** coil, **13** turns, **14** turns, **15** voltage, **16** volts **17** distribution, **18** supply

Situations: Phrases for presentations

1. I'm *glad* to have the chance to speak to you about ...
2. I *appreciate* that you're sacrificing your valuable time to listen to me.
3. Don't hesitate to *interrupt* me when you have a question.
4. The *topic* of this presentation is ...
5. Let me first give you an *overview* of the situation.
6. My presentation will be in three *main parts*.
7. I'm going to explain the problems *involved* in ...
8. May I draw your *attention* to the handout in front of you?
9. So to *summarise* we can say that ...
10. The *upper part* of the slide shows the interior of the container.

Beef up your word power

1. What do you get when you assemble the components?

transformer	core, voltage, coil
generator	iron core, coil, magnet
power plant	chute, turbine, generator
nuclear reactor	control rods, uranium pellets, coolant pump

2. Words around a nuclear power plant

Match definitions	and parts of the PWR
1 It contains rods of uranium.	B reactor core
2 They stop the chain-reaction.	C control rods
3 It should survive most accidents intact.	D containment
4 It prevents water from turning to steam.	E pipes
5 They prevent the overheating of the core.	A water circuits
6 They take high pressure stream to the turbine.	F pressuriser

Our quiz for the experts

Are these incredible statements true or false?

1. Large-scale steel production started in the USA after the steel converter was invented by William Kelly. **T**
2. Gotlieb Daimler fitted his first petrol engine to a boat in 1873. **T**
3. He fitted his second motor to a bicycle in 1875. **T**
4. The first steam engine invented by Hero was so impractical that it vanished until the 1700s. **T**
5. The British inventor Joseph Wolson Swan invented the light bulb in 1878, one year earlier than Edison. **F**
6. The largest blades measure up to 100 meters. **F**
7. The nuclear reaction that happens inside a reactor is called nuclear fission (not fusion), that is an atom breaks apart under the impact of a neutron. **F**
8. A nuclear reactor doesn't convert energy, it creates energy by changing mass into energy. **F**
9. PWR is short for pressurised water reactor. **T**
10. The control rods in a reactor are made of graphite. **F**

CHAPTER 7

Working with the text

1. Are these statements true (T) or false (F)?

1.	At Ruver they have internal E-mail.	**F**
2.	In Germany a cellular phone is called 'handy'.	**T**
3.	Extensions are the phones in the company.	**T**
4.	On the dial you key the telephone number in.	**F**
5.	A host is a main-frame computer.	**T**
6.	At Ruver the sales force can access the host computer.	**F**

2. What James might have said but didn't

1.	The *handset* is the part that we lift when we want to make a call.	The *receiver* is the part that we lift when we want to make a call.
2.	The various *telephone companies* have got the *solutions.*	The various *providers* have got the *answers.*
3.	*With the help* of this medium we can exchange ideas.	*By means* of this medium we can exchange ideas.
4.	I think we'd better leave that *subject.*	Well, I think we'd better leave that *theme.*

How things work in the phone

1. Are you a technical genius?

(**1**) handset (**2**) mouthpiece (**3**) loudspeaker (**4**) earpiece (**5**) microphone (**6**) sound waves (**7**) diaphragm (**8**) resistance (**9**) carbon granules (**10**) wires (**11**) exchange (**12**) electromagnet (**13**) coils (**14**) diaphragm

2. Things are getting even more complex

(**1**) electrical signals (**2**) fibre-optic cables (**3**) microwave links (**4**) electrical signals (**5**) multiplexer (**6**) decoder (**7**) demultiplexer

3. What is the 'it'?

It is the part of the phone you hold.	handset
It is a thin metal disk that can vibrate.	diaphragm
It is a grain of special coal.	carbon granule
It changes sound into electrical current.	microphone/mouthpiece
It encodes digital signals into binary.	multiplexer
It changes electrical current into sound.	loudspeaker/ earpiece
It directs the signals to the right line.	exchange

Situations

In a public telephone box

1. Lift the receiver.
2. Insert your money.
3. A minimum of 10 p is required.
4. Listen for the dial tone.
5. Dial the number required.
6. Do not insert money. This call is free.
7. Emergency calls: dial 999.
8. First dial the international code,
9. ... then the area code minus initial '0'... ,
10. ...then the customer's number.

Beef up your word-power

1. What do you need to assemble to get these parts of the phone?

decoder
ISDN system — multiplexer
demultiplexer

case
telephone — handset
dial

carbon granules
microphone — electromagnet
diaphragm

2. Do you know the difference between old and new phones?

Old phones	New phones
They have rotary dials.	They have keyboards.
They have cords.	They are cordless.
Sound waves are converted into electrical signals.	Sound waves are converted into light signals.
They travel through coaxial copper cables.	They travel through optical-fibre cables.

3. Do you know your fax machine?

(**1**)scanning device (**2**) pixels (**3**) electrical signals (**4**) telephone line (**5**) receiving machine

Q The quiz for our experts

1. Alexander Graham Bell invented the telephone in 1796.
2. Bell had spilt acid from a battery down his trousers and needed help from his assistant. He cried: "Mr Watson. Come at once. I want you". Watson, who was in a room next door, could hear the words over the newly invented phone.
3. It is wrong. Alexander Graham Bell invented both the loudspeaker and the microphone.
4. The telephone signals in optical-fibre cables are produced by a (**A**) semiconductor laser.
5. A pair of individual fibres can carry as many as 2,000 calls (**B**) simultaneously.
6. ISDN stands for 'Integrated Services Digital Network'.

C Let'sdo it crosswise

				1 I			2 D	3 E	C	O	4 D	E
	5 M	A	G	N	E	T		X			I	
	O			S				P		6 F	A	X
	7 U	R	G	E	N	8 T		E			L	
	T			R		A		R		9 R		
	H			T		10 P	11 A	T	T	E	R	12 N
	P						N			C		O
13 W	I	T	T	Y			A			E		D
	E				14 B	U	L	B		I		
	C				L		O			V		
	15 E	X	C	H	A	N	G	E		E		
					D		U			16 R	O	D
17 C	R	A	D	L	E		E					

CHAPTER 8

T Working with the text

1. Are these statements true (T) or false (F)?

1.	James Kelly has bought an MBA car.	F
2.	Dr. Block called James to compliment him on his trip to Indonesia.	F
3.	Dr. Block has announced major changes at Ruver.	T
4.	Ruver has a LAN and facilities for video conferences.	F
5.	With a laser pointer you switch the TV on and off.	F
6.	In the 'pastor position' you stand with your hands in your pockets.	F
7.	Obviously there are rumours about a German take-over.	T

2. What they might have said but didn't

1. What are you *driving* at?	But what are you *getting* at?
2. You may *have* to become more of a manager.	You may *be required* to become more of a manager.
3. *You win some, you lose some.*	It's a question *of ups and downs* in Ruver.
4. *The important* things first.	*First* things first.
5. We've certainly a few things to *sort out.*	We've certainly a few things to *tidy up.*
6. I get the *message.*	I get the *point.*

H How things work in the technical office

1. Which medium would you choose for ... ?

B	an informal training meeting	3 – 5	participants
C	discussion of projects	5 – 10	participants
A	selling office space	30 – 150	participants
D	introducing new sports equipment	20 – 50	participants

2. Toolbox for the secretary: What is 'it'?

It hurts when you sit on it.	drawing pin
It is not as safe as a stapler.	paper clip
It is empty when you need it most.	stapler
It makes holes to fit the folder.	punch
It usually sticks to your fingers.	sellotape
It is usually dried up when you need it.	liquid paper

S Situations

1. Check whether you got the message

If I've *understood you correctly*, you are saying that we should give up the whole project?
Would *I be right* in saying that you are against the new model range?
Am I *correct in assuming* that you have a better plan?
Correct me *if I'm wrong*, but aren't these figures those from last year?

2. Introduce a question

May I ask where you got these figures from?
Would you mind if I asked you a personal question?
I wonder if I could ask you what you think of the project ?

3. Ask for explanation

I'm afraid we weren't quite *able to catch* what you were saying.
I'm afraid I didn't *quite follow* what you said about ...
I'm afraid *I'm not quite clear* what you mean by that.
Sorry, I didn't *get your last point.* Could you go over it again, please?

W Beef up your word-power

1. What do you get when you assemble the components?

felt pen/tip	housing, refill, cap
desk calculator	keys, display, case
swivel chair	seat, back, stand
dictaphone	microphone, loudspeaker, tape

2. Lines on your transparency

This (______) is a continuous line. This (.......) is a dotted line.

This (_ _ _ _ _) is a broken / dashed line.

3. Shapes for your impact sheets

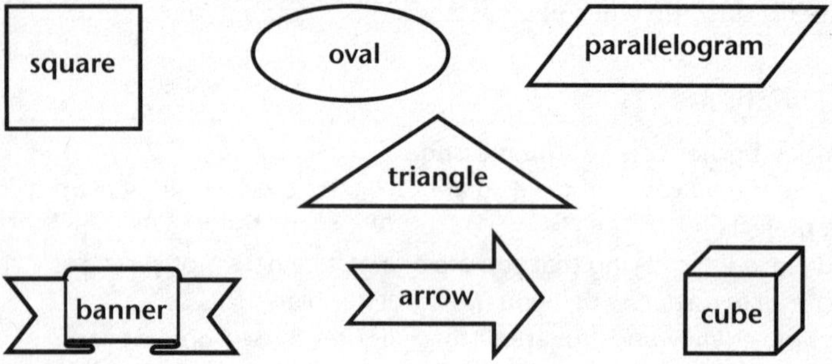

CHAPTER 9

Working with the text

1. Are these statements true (T) or false (F)?

1.	The department consumes about 2,000 copies a month.	F
2.	James used to work for a manufacturer of copiers.	T
3.	If the cover flap is open you use ten to twenty times more toner.	T
4.	The department uses recycled paper.	T
5.	James recommends buying a printing press.	F
6.	James suggested buying Gondex copiers.	F

How things work : Getting to know your copier

Toolbox 1

document glass	2
document cover	1
operational panel	3
front cover	4
handle	6
toner/developer cartridge	5
paper feed slot cover	8
paper feed slot	7
paper guide	9

Toolbox 2

release button	1
power cord socket	3
power switch	2

Toolbox 3

photo-conductive drum	2
fusing area	1
bottom cover	4
drum cartridge release lever	3

Toolbox 4:

key/indicator	nr	key/indicator	nr
reduction /enlargement	2	stop/cancel key	5
copy quantity	4	exposure key	1
misfeed indicator	3	start/ready	6

Situations

Don't you think it might be a good idea to plug the copier in first?
Wouldn't it be wiser to wait unil the photoconductive drum has cooled down?
If I may make a suggestion, always shut the document cover before copying.
Wouldn't *it be better* to buy a series of new generation copiers?
You'd better use quality paper on this copier.

Why not take a more efficient copier?
What about buying the Gondex Copy Miracle?
I would suggest that we put off the decision.

W Beef up your word power

1. What is the 'it?'

It contains black powder.	toner/developer cartridge
It makes copies smaller.	reduction key
It is a key to blow copies up.	enlargement key
It stops the copying process.	cancel key
It blinks when paper jams.	misfeed indicator
It switches the electricity off.	power switch

2. Making a copy

1. *Turn* the power switch on.
2. *Open* the document cover.
3. *Place* the document on the document glass.
4. *Choose* the exposure for light or dark originals.
5. *Insert* a sheet of paper.
6. *Press* the start key.
7. *Remove* the copy.
8. *Switch* off the power.

C Let's do it crosswise

			1 T			2 A	D	V	3 I	S	E	
4 P			A		5 S				N			
L			K		P		6 S		S			7 C
8 A	P	P	E	A	L		9 O	P	E	N		O
C					I		U		R			P
E		10 F	A	11 S	T	E	N		12 T	I	D	Y
		I		E			D					
		D		T		13 A					14 R	
		D			15 P	R	O	M	O	T	E	
		L		16 J		G					M	
		17 E	V	A	L	U	A	T	E		O	
				C		E					V	
	18 S	H	A	K	E		19 C	A	N	C	E	L

CHAPTER 10

Working with the text

1. Are these statements true (T) or false (F)

1.	Ruver have bought the majority interest in AMW.	**F**
2.	AMW manufacture up-market off-road cars.	**F**
3.	AMW promised not to reduce the number of employees.	**T**
4.	AMW will negotiate salaries with the trade unions.	**F**
5.	Those who don't reach the target will be fired.	**T**
6.	Steve will be involved in the rationalisation process.	**T**
7.	The Japanese will manufacture the engines for Ruver.	**F**
8.	James is getting on well with the Japanese.	**T**

2. What they might have said but didn't

1.	Aren't the two companies *competitors?*	Aren't the two companies *in competition* with each other?
2.	Will you be involved in the *reorganisation?*	Will you be involved in the *restructuring?*
3.	I have to identify *parallel* processes.	I have to identify *redundant* processes.
4.	You'd better *keep out of the way*	You'd better *keep your head down*.
5.	Why are our eastern friends *unpopular* with AMW?	Why are our eastern friends *in AMW's bad books?*
6.	I see what you're *driving* at.	I see what you're *getting* at.

C Let's do it cosswise

1 F								2 W			3 B	
4 A	S	5 S	U	6 M	E		7 L	A	Y	E	R	
I		K		A				L			I	
N		I		8 S	U	9 P	P	L	Y		C	
T		M		O		R					K	
	10 R		11 A	N	N	O	U	N	C	E		
	E		S			V					12 I	
	13 P	O	S	I	T	I	O	N			N	
	L		E			D			14 S	I	T	E
	A		M		15 L	E	V	E	L		E	
	C		B						O		N	
	E		L						P		D	
		16 D	E	C	O	R	A	T	E			

CHAPTER 11

T Working with the text

1. Are these statements true (T) or false (F)?

1.	James is in Munich.	**T**
2.	James had forgotten his pass.	**F**
3.	AMW do not want to buy from Japanese suppliers.	**T**
4.	Printing presses cheaper are for short runs.	**F**
5.	Herr Sammuk proposes a letterpress.	**F**
6.	James thinks that the printing press is similar to a colour copier.	**T**

2. Words from Chairman Kurz

1.	Ladies and gentlemen, *may I open this meeting* now?	Ladies and Gentlemen, I *declare this meeting open.*
2.	The *subject* of our meeting is Mr Kelly's research on ...	The *purpose* of today's meeting is Mr Kelly's research on ...
3.	*Now to the matter.*	*Let's get down to business.*
4.	Is there *anything else to discuss*?	Is there *any other business*?
5.	That *concludes our business for today.*	I *declare the meeting closed.*

H How things work: Printing press

3. The rotary printing press in five confusing steps

1	The films are	E	fitted round printing cylinders.
2	The printing plates are	C	wet in the dampening unit.
3	Sheets of paper are	A	passed through the press.
4	The four colours are	B	printed one after the other.
5	The sheets are first printed	D	on one side, then on the other.

S Situations: The language of meetings

1. The *reason* for our coming together is to ...
2. Let' get down to *business*, shall we?
3. Would anybody wish to add an *item* to our agenda?
4. There are three *problems* that should be solved today.
5. The first item on the *agenda* is ...
6. May I ask for *reactions* to Mr Kelly's suggestions?
7. Any further *points* before we bring this discussion to an end?
8. Let's now turn to the *problem* of pollution.
9. To sum up, it seems that we have a *consensus*. We´re all agreed.
10. Could we *adjourn* the meeting to Monday?

2. A guided translation

1.	Eröffnung:	*Ladies and* Gentlemen, I *declare* the meeting *open*.
2.	Begrüßung:	I'm happy to *welcome all* of you *to* our meeting.
3.	Verspätung:	May I *apologise* for the *fact* that Mr Kelly is late.
4.	Ergänzung:	I'd like to *add* an *item* to the agenda.
5.	Zweck:	The *purpose* of today's meeting is ...
6.	Tagesordnung:	As you can *see from* the *agenda* we must ...
7.	Resümee:	If I *could* just *sum* up the discussion.
8.	Beenden:	Well, I *think* we've *discussed* everything.
9.	Schließen:	If *nobody* has anything to *add* I'd like to *draw* the meeting to an *end*.

Beef up your word power

1. What is the 'it'?

It separates the colours.	colour scanner
It is a coloured liquid.	ink
It wets the printing plate.	dampening unit
It covers the film.	light sensitive coating
It has a film wrapped around it.	cylinder
It rejects water	lacquer/coating

Our quiz for laymen and experts

1. Which other raw material was used in Germany before World War II? (**B**) hemp
2. Where did people first write on wax tablets? In Greece
3. In 105 AD paper was invented. In China.
4. Which people brought the paper to Europe? (**C**) the Arabs
5. Where was the first book produced in 868? (**A**) in China
6. The first movable types were invented (**A**) in China in 1041.
7. The first movable types were made of (**B**) clay
8. The printing press with movable metal type was invented in (**C**) 1450.
9. Johan Gutenberg invented the first printing press with movable metal type.

Let's do it crosswise

	[1] T	H	O	R	O	U	G	H		[2] S		
	A									U		
	[3] S	T	[4] I	C	K	E	[5] R			B		
	K		R				E			J		
		[6] J	O	B			[7] S	H	E	E	T	
			N				E			C		[8] P
	[9] M				[10] D	I	A	M	E	T	E	R
	[11] I	S	S	U	E		R					E
	S				V		C		[12] F	[13] I	L	M
	H				I		H			N		I
[14] H	E	M	P		C					K		S
	A			[15] R	E	A	S	O	N			E
	R											S

CHAPTER 12

Working with the text

1. Did you get the message?

1. What is Sheila's background? (**C**) She designed underwear for a department store.
2. What are James' present activities at Ruver? (**B**) He is training students in electronic data processing.
3. Information travels in your computer in (**C**) bytes.
4. What is a byte? It is an information unit consisting of 8 bits.

2. What they might have said but didn't

1. If a software is *fool-proof*, only a fool will use it.	If a software is *idiot-proof*, only an idiot will use it.
2. We don't *have to* know much more in theory.	We don't *need to* know much more in theory.
3. I *have a degree* in arts and applied psychology.	I *graduated* in arts and applied psychology.
4. A computer is actually a *simple* device.	A computer is actually a *straight-forward* device.
5. Listen Sheila, what do you say if we stop here?	Listen Sheila, what do you say if *we call it a day*?

How things work in the computer

Let's put your brain to the test

You have got	What about your computer?
1 short-term memory	Random-Access-Memory
2 long-term memory	Read-Only-Memory
3 five senses	input-units
4 knowledge and skills	programs
5 pen, agenda, notepad	output units
6 parents and teachers	programmers
7 grey cells	chips

Ⓢ Situations

Working with your computer

1. First *switch on* the computer.
2. *Call up* your word processor by clicking on the right icon.
3. *Open* a new file.
4. *Key* in the name of the new file so that you can *retrieve* it again.
5. Write your document and *format* it so that it looks good.
6. Correct and *edit* it, if necessary.
7. Don't forget to *save* it.
8. Switch on your printer and *print it out*.
9. *Change* to drive A and make a safety copy on a floppy.

Ⓦ Beef up your word power

1. What do you get when you assemble the components?

	disc
floppy	guard
	window

	housing
monitor	screen
	switch

	arms
hard disc	heads
	discs

	circuits
chip	silicon base
	photoresist

2. What is the 'it'?

It is the code the computer understands.	binary code
It is the general name for information.	data
It is an output unit that displays the data.	monitor
It is the information that comes out on paper.	printout
It is the computer's 'short-term' memory.	RAM
It is the computer's 'long-term' memory.	ROM
It is a portable magnetic disc.	floppy
It is a tiny electronic component.	chip

Let's do it crosswise

				1 P	E	2 R	S	O	N	3 A	L	
				O		A				B		4 M
	5 T			6 R	O	M		7 S	T	O	R	E
	8 E	D	I	T				O		R		M
	M			A				F		T		O
	P			9 B	U	10 S		T				R
	O			L		U		W		11 K	E	Y
12 D	R	I	V	E		P		A				
	A					P		13 R	A	14 T	E	
15 C	R	A	16 S	H		O		E		Y		
	Y		A			R				P		
			V			T		17 S	P	E	E	D
18 F	I	L	E									

CHAPTER 13

Working with the text

Are these statements true (T) or false (F)?

1.	They are driving in an AMW car.	**T**
2.	Mr McDray is specialised in breeding various sorts of animals.	**F**
3.	James has Irish forefathers.	**T**
4.	Sheila is looking for a husband to take over the farm.	**F**
5.	Mr McDray needs the peripherals to make maps.	**F**
6.	A plotter is a sophisticated printer for graphics.	**T**

H How things work

Put the numbers from the tool box next to the arrows

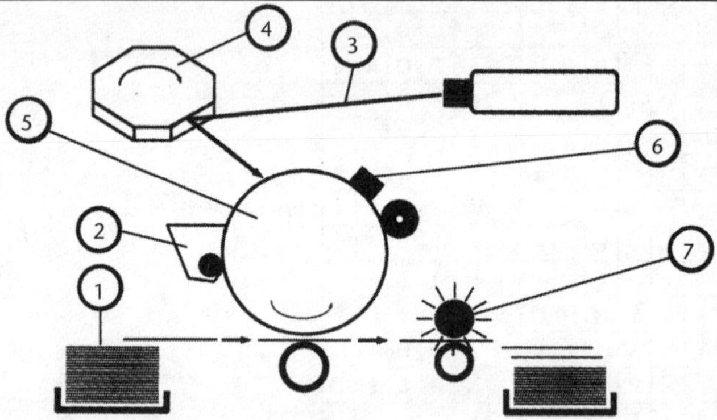

W Beef up your word power

1. Odd man out

scanner	ASCII	heater	spinning mirror
printer	phoneme	toner cartridge	metal drum
computer	grapheme	laser beam	electrical charges
laser	syllable	spinning mirror	fusing system

translate	control	store	amplify the beam
arrange	check	save	focus the beam
convert	verify	file	convert into ASCII
transfer	correct	retrieve	build up an image

Intelligent guessing: What is the 'it'?

It converts computer signals into sound.	synthesiser
It focuses beams of light.	lens
It is a computer code	ASCII
It recognises shapes	scanner
It is a component that seals toner to paper	fusing system
It is a super printer for graphics.	plotter

Q The quiz for super experts

1. ASCII means **A**merican **S**tandard **C**ode for **I**nformation **I**nterchange.
2. The letters OCR stand for **O**ptical **C**haracter **R**ecognition.
3. Where does the word laser come from? **L**ight **A**mplification by **S**timulated **E**missions of **R**adiation
4. What is the device called that bends light rays to form an image? It is a lens.
5. The Roman Emperor Nero, AD 37 was one of the first people to use a lens when he watched performances in the arena. He looked through a fragment of emerald to improve his poor eyesight.
6. All rays of ordinary light have the same wavelength. Is this true or false? It's false. The rays of laser light have the same wavelength and frequency.
7. A kilobyte is (**B**) 1024 bytes.
8. What is a 'cog'? It is (**C**) toothed gear wheel.
9. What is the point called at which rays or waves meet? The focus
10. Which device splits up the image into a series of lines and converts the various of brightness into electrical signals? A scanner, of course.

C Let's do it crosswise

			[1]C	H	[2]A	R	[3]A	C	T	E	R	
[4]S	I	Z	E		R		M					
T			R		T		P				[5]D	
A			E				[6]L	A	[7]S	E	R	
T		[8]L	A	Z	Y		I		E		U	
E			L				F		A		M	
	[9]L	A	S	[10]S	Y		Y		L			
	E			C				[11]D		[12]S		[13]N
	N		[14]F	R	E	Q	U	E	N	T		E
	S			A				P		[15]O	C	R
		[16]P	I	T	C	[17]H		T		R		O
				C		I		H		E		
[18]R	O	W	O	H	L	T						

CHAPTER 14

T Working with the text

1. Are these statements true (T) or false (F)?

1.	Ruver cars will be fitted with LOOF.	F
2.	The GPS has been developed by Japanese military.	F
3.	Making use of the old boys' network means using connections.	T
4.	James is sent to Japan because he has connections there.	T
5.	The computerised tram gets its signals from satellites.	F
6.	Mr Kopp wants to develop a driverless Ruver car.	F
7.	The Japanese are developing a navigation system on the basis of GPS.	T

2. What they might have said but didn't

1.	We *mustn't fall behind* our schedules.	We *must keep to* our schedules.
2.	There are cameras at the front and *the back.*	There are cameras at the front and *rear.*
3.	The pictures are *transmitted* to the controller.	The pictures are *relayed* to the controller.
4.	We must *be up to date.*	We must be *at the state of the art.*
5.	The Japanese are our only real *rivals.*	The Japanese are our only real *competitors.*
6.	Quality *is expensive.*	Quality *has its price.*

H How things work

1. Tools in use: At the electronic bus stop

(1) electronic map, (2) pay and change machine (3) electronic map (4) pay and change machine (5) credit card slot (6) flap (7) ticket slot

2. Tools in use: Understanding the driveless bus

A 3, B 2, C 1, D 4, E 5

S Situations: What to say when things don't work

1. Expressing doubt and uncertainty

1. *Are you sure* we should be relying on this system completely?
2. Well, it *seems to me* that it's taking us through some minor roads.
3. *What if the* Ruver navigator got its figures all wrong?
4. Buxton must be on the other side of the field – I wouldn't *bet on* it
5. Why are we going slower? – I can't say *for certain.*
6. I haven't the foggiest *notion.*
7. What are we going to do? – I haven't *a clue.*
8. Well, I wouldn't *rely on it.*

2. Don't worry!

Es gibt keinen Grund zur Aufregung.	There's no reason to be alarmed.
Du brauchst dir keine Sorgen zu machen.	There's nothing to worry about.
Sieh nicht immer alles so schwarz.	Don't always look on the black side.
Alles hat auch eine gute Seite.	Every cloud has a silver lining.

C Let's do it crosswise

1 A	2 R	T				3 R		4 B		5 R		
	U		6 S	C	H	E	D	U	L	E		7 A
8 F	L	A	P			A		S		L		L
	E		9 E	N	T	R	Y		10 L	A	N	E
			C					11 D		Y		R
12 P	O	13 S	S	E	S	14 S	O	R				T
R		C				K		I		15 T		
O		H			16 S	I	L	V	E	R		
U		E		17 O		R		E		18 A	I	19 M
D		20 M	O	N	I	T	21 O	R		C		I
		E					C			K		N
			22 M	A	J	O	R					O
										23 F	O	R

CHAPTER 15

T Working with the text

1. Are these statements true (T) or false (F)?

1.	An extra large fan muffles the noise.	T
2.	Hot gases are forced out of the back at high speed. That propels the plane through the air.	T
3.	Jet engines are not combustion engines.	F
4.	Birds cannot bring down a jet plane.	F

2. What they might have said but didn't

Let's shake hands.	Put *it there.*
Pleased to meet you.	*Glad to* know you.
It's more complex than that.	It's not *as simple as* that.
Let's be serious.	But *joking* apart.
We are absolutely safe.	We're as *safe as houses.*

3. A little grammar for a change

1. If God *had wanted* men to fly, they *would have been* born with wings.
2. I *wouldn't be* very happy flying in a car, either.
3. There *could be* something wrong with components *supplied.*

H How things work in the plane

1. Verbs in action: The jet engine

The engine that *propels* big airliners *sucks* air in and *ejects* it at high speed. At the front of the engine, at the intake, a large fan *rotates* to *draw* air in. Some of the air *cools* the engine. The mainstream *enters* the compressors, which *contain* both rotating and stationary blades. The compressors *increase* the pressure of the air, which then *flows* to the combustion chamber where it is *mixed* with kerosene or paraffin. The mixture *burns* explosively and *expands.* The hot, high-pressure air *speeds* towards the exhaust with tremendous force and *pushes* the plane through the air. On its way the hot gas *passes* through turbines that *drive* the fan and the compressors.

2. Put the right number into the circles

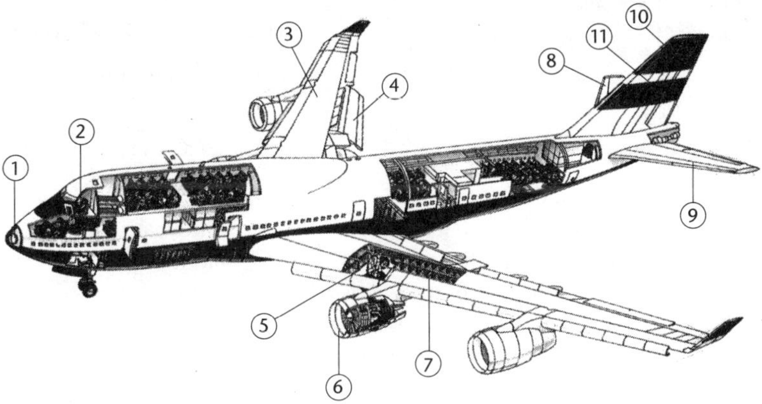

Situations: At the airport

1 A check-in desk, **2 A** book an early flight, **3 B** gate do I depart, **4 D** the departure lounge, **5 B** duty-free allowances, **6 A** items of luggage, **7 A** leave / unattended, **8 C** proceed to, **9 A** boarding at, **10 D** last call

Beef up your word power

1. One verb is different. Mark it!

enter	propel	contain	flow
eject	push	mix	speed
suck in	drive	burn	heat
take in	rotate	heat	pass

2. Three parts are in the same area, one is not. Mark it!

fan	rudder	tail fin
turbine	fuel tanks	elevator
compressor	engine	radar
undercarriage	aileron	rudder

3. Where do these components belong?

wing	aileron, tanks, engine
engine	fan, compressor, turbine

4. What is 'it' used for?

It causes the airliner to fly a curve.	rudder
It is needed during take-off and touch-down.	undercarriage
It is the part reserved for the pilots.	flight deck
It can enlarge the surface of the wings.	aileron
It causes the aircraft to gain or lose height.	elevator
It has a curved surface and produces the lift as it moves through the air.	wing

The quiz for our expert

1. (**A**) The explorer, Marco Polo, reports that in China in the 12th century criminals were lifted by large kites as a punishment. Kite flying was also used to look out over enemy territory.
2. (**A**)The first balloon to carry passengers was built by the French brothers Montgolfier in 1783. They used the first gas-fired balloon that carried them over a distance of 8 kilometres.
3. Is not true: Jacques Charles piloted the first balloon that was steam-powered in 1852. Its inventor was the French engineer, Henri Griffard.
4. (**C**) The principle of the curved surface of the wing (aerofoil) produces the lift that keeps the plane in the air. This principle was discovered by Daniel Bernouilli in 1738.
5. (**B**) The first use of the principle of the wing (airfoil) was in ships. In Britain in 1861 Thomas Moey tested wings by fixing them beneath a boat and found that they raised it above the water. It was further developed in Italy.
6. (**B**) The Wright brothers flew first powered aeroplane at Kitty Hawk in North Carolina in 1903. Unlike all modern aircraft it had no ailerons. They were developed by the British engineer Henry Farman in 1908.

7. The first helicopter, invented by Paul Cornu, whirled through the air in 1907. The development of a reliable helicopter came about thirty years later.

C Let's do it crosswise

		[1] F	L	[2] I	G	H	T			[3] B		
		L		N				[4] P		A		[5] A
		I		T			[6] W	I	N	G		I
[7] S	T	E	W	A	R	D		L		G		R
U		R		K				O		A		C
C				[8] E	J	E	[9] C	T		G		R
[10] K	I	[11] T	E				R			E		A
		U		[12] G	A	T	E		[13] F			F
		R					[14] W	R	I	G	H	T
		[15] B	L	A	[16] D	E			N			
		I			R							
		N			[17] A	I	R	P	L	A	N	E
		E			W							

Anmerkung: Airplane (= amerikanisch)
aeroplane (= britisch)

Appendix 1: General tools – Terms & phrases

accessories	Zubehör
accessories included in the price	Zubehör im Preis inbegriffen
accumulator (battery)	Akkumulator, Batterie
charge up an accumulator	eine Batterie aufladen
adjustable spanner (wrench)	verstellbarer Schraubenschlüssel
tighten a nut	eine Mutter anziehen
aerial	Antenne
put up an aerial	eine Antenne aufbauen
transformer	Transformator
air cooled transfomers convert the current	luftgekühlte Transformatoren spannen den Strom um
Allen key	Imbus-Schlüssel
alternator; it generates electricity	Wechselstromgenerator; er erzeugt Elektrizität
anemometer; it measures the wind speed	Windmesser; er mißt die Windstärke
angle	Winkel
awl; punch / make holes	Ahle; Löcher mit einer ~ stechen
bearing; it reduces friction	Lager; es vermindert Reibung
journal bearing	Gleitlager
ball bearing;	Kugellager
bench vice	Schraubstock
blast furnace; it melts ores	Hochofen; er schmelzt Erze ein
blower	Gebläse
bolt; you tighten a ~	Bolzen; man zieht einen ~ an
plain bolt	einfacher Bolzen
threaded bolt	Gewindebolzen
boiler; heat up the ~	Kessel; den ~ anheizen
brace	Spannriegel, Spreize
boring tool; ~ make (drill) holes	Bohrwerkzeug; Bohrwerkzeuge bohren Löcher
brake system	Bremssystem; die Bremsen
the brakes stop the engine	stoppen die Maschine
brush; you paint with a ~	Pinsel; man streicht mit einem ~

cable	Kabel
lay the cables	Kabel verlegen
single-core cable	Einleiter-Kabel
cable-cutter	Kabelschneider
conduit; you lay cables in a ~	Rohrleitung; man verlegt Kabel in eine ~
calipers	Taster, Tastlehre, Tastzirkel
outside calipers	Außentaster
inside calipers	Innentaster
check the distance with ~	den Abstand mit einem ~ prüfen
camshaft	Nockenwelle
capacitors	Kondensatoren
carburettor; it mixes fuel with air	Vergaser; er mischt Kraftstoff mit Luft
cells	Zellen
chisel	Meißel
chuck	Einspannvorrrichtung
circuit diagrams	Schaltpläne
clutch; press the ~	Kupplung; die ~ drücken
cog-wheel; is part of a gear	Zahnrad; es ist Teil eines Getriebes
combustion engine; it burns petrol or diesel	Verbrennungsmotor; er verbrennt Benzin oder Diesel
compressor; it builds up compressed air	Kompressor; er verdichtet Luft
control panel	Steuertafel
control console; you control operations from a ~ ~	Steuerpult; man kontrolliert Vorgänge von einem ~ aus
conveyor belt; it moves the parts along	Förderband; es befördert Teile
counter sink; you counter sink a screw with it	Senkbohrer; man versenkt Schrauben damit
crank shaft	Kurbelwelle
cranes lift things up	Kräne heben Gegenstände hoch
tower crane	Hochkran
gentry crane	Portalkran
wharf crane	Hafenkran
depth gauge checks the depth	Tiefen-Meßschieber prüft die Tiefe

dowel	Dübel
drilling machine	Bohrmaschine
drum	Trommel
dynamo creates energy and light	Lichtmaschine erzeugt Energie und Licht
fan	Lüfter
fanbelt	Riemen
file; you file things down	Feile; man feilt Gegenstände ab
flat file	Flachfeile
square file	Vierkantfeile
round file	Rundfeile
first-aid kit	Erste-Hilfe-Kasten
fittings; attach ~	Beschlagteile anbringen
fork-lift truck; it lifts pallets	Gabelstapler; er hebt Paletten
fuse;	Sicherung
check the fuse	eine Sicherung überprüfen
a fuse has blown	eine Sicherung ist durchgebrannt
G-clamp; it clamps wood together	Schraubzwinge; sie preßt Holz aufeinander
gauge; it measures distances	Lehre; sie mißt Abstände
gear-box;	Getriebe
gear-box has crashed	das Getriebe hat sich gefressen
gears; ~ mesh together	Zahnräder greifen ineinander
spur gear	Stirnradgetriebe
helical spur gear	Stirnradgetriebe mit Spiralverzahnung
worm gear	Schneckenradgetriebe
goggles; sometimes they are fogged	Schutzbrille; manchmal sind sie beschlagen
glue; ~ is hard	Klebstoff; ~ ist ausgetrocknet
saw	Säge
cuts boards in half	Bretter entzwei sägen
hack-saw	Bügelsäge
hammer	Hammer
hammer a nail in	einen Nagel einschlagen
ball hammer	Rundhammer

claw hammer	Klauenhammer
drill	Bohrer
hand drill	Handbohrer
wood drill	Holzbohrer
hoist	Lastenaufzug; Winde
hoist pulley	Seilrolle
housing	Gehäuse
jack	Heber
wheeled car jack	fahrbarer Wagenheber
jack up a car	ein Auto aufbocken
joint; it fixes two things together	Verbindung; sie verbindet zwei Dinge
jib	Ausleger
ladder; to put up a ~	Leiter; eine ~ anlegen, aufstellen
lathe	Drehbank
lifting gear	Hebevorrichtung
machining	Maschinenbearbeitung
drilling	Bohren
grinding	Schleifen
turning	Drehen
milling	Fräsen
mallet	Fäustel
mason's mallet	Steinhauer-Fäustel
mandrel	Dorn
mason	Maurer
metal shears	Metallschere
moulding	Former
mould a part	Ein Teil gießen
pouring	Gießen
casting	Gußstück
paint	Farbe
pawl	Sperrklinke
pliers	Zange
pull nails out	Nägel herausziehen
plane; to plane a board	Hobel; ein Brett hobeln
pressure gauge; check the pressure	Manometer; den Druck prüfen

prick punch	Schlagdorn
pulley	Flaschenzug, Rolle
punch	Schlagstempel
scaffolding;	Gerüst; Staffelei
put up the ~	das ~ aufstellen
scraper	Spachtel
screw	Schraube
screw a screw in	eine Schraube eindrehen
screw with hexagon head	Sechskantkopf
screw with countersunk head	Senkkopf
screwdriver	Schraubenzieher
tighten screws	Schrauben anziehen
soldering iron	Lötkolben
solder	löten
spanner	Schraubenschlüssel
ring spanner	Ringschlüssel
box spanner	Hohlschlüssel
spider	Kreuzschlüssel
spirit level	Wasserwaage
lay a ~ ~	eine Wasserwaage anlegen
trowel	Kelle
spread mortar with a trowel	Mörtel mit der Kelle verteilen
plastering trowel	Verputzkelle
smooth plaster with a plastering trowel	Gips mit einer Verputzkelle glattstreichen
union socket	Verbindungsmuffe
a ~ joins parts	eine ~ verbindet Teile
vice	Schraubstock
put in the vice	einspannen
bench vice	Werkbank-Schraubstock
washer	Beilagscheibe
put a ~ under the screw	eine ~ unter die Schraube
legen	
flat washer	flache Unterlegscheibe
lock washer	Federring

winch	Winde
to winch something up	etwas hochwinden
wire	Draht
wrench	Schraubenschlüssel
loosen the nuts with a wrench	Muttern mit einem ~ lockern
ratchet wrench	Engländer
monkey wrench	Franzose
screw wrench	Universalschlüssel

Appendix 2: Speaking about graphs & charts

When we talk about sales, results or figures and statistics in industry we often use the visual form – the graph.

The graph explains at a glance what a given situation is. There are different types of graphs:

Bar graph or chart:

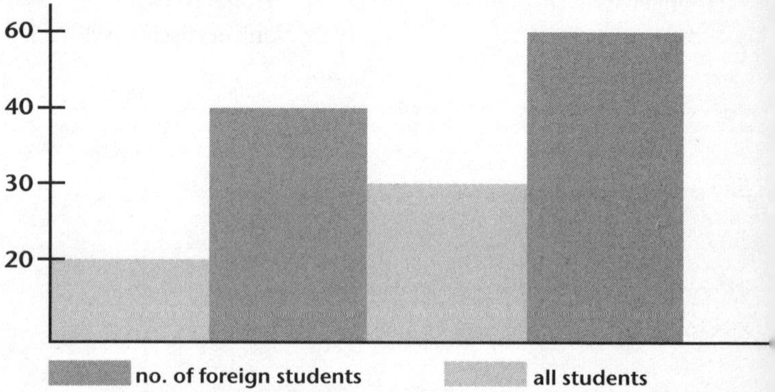

Pie chart or graph:

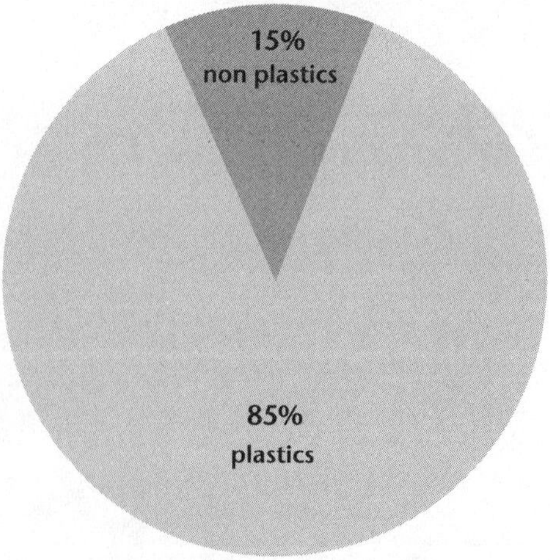

Line graph

We use a line-graph to give various sorts of information. It is especially common in sales and marketing. The different types of lines are as follows:

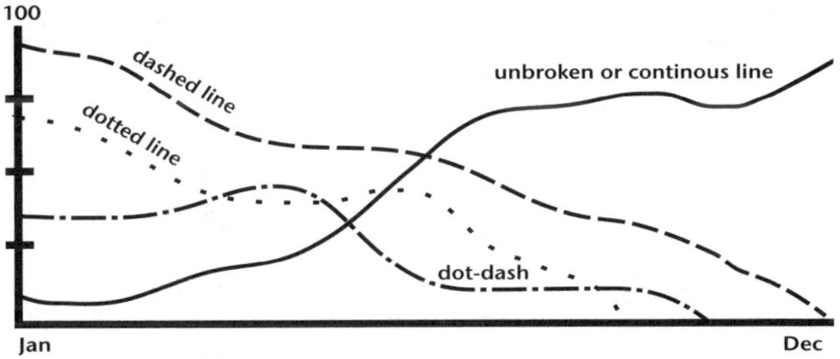

The dotted line represents the sales of tape-recorders. The unbroken lines show the development of costs in the purchasing department.

We use a special "graph" language to describe what is happening on the lines. This describes the facts that the lines show.

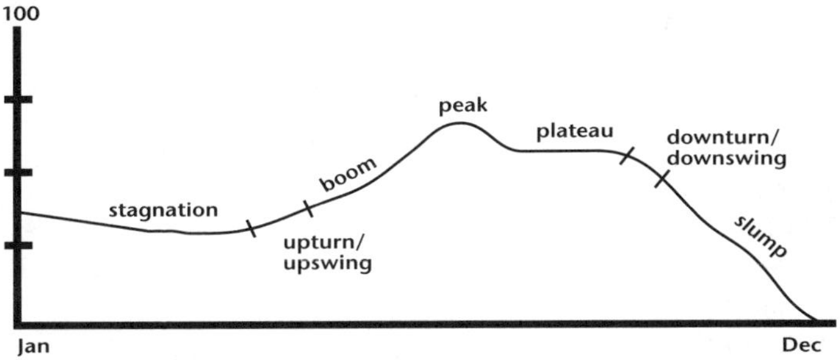

For example, regarding sales or production we could say:

an upturn/upswing in production will mean that ..
a boom in printers will help to overcome ...
a peak in sales can be expected at ...
a plateau in investment was the result.
a downturn/downswing in funds means that we cannot ...
a slump in orders will kill the company.

There are some typical words that we use to describe graphs. They depend on direction:

Direction

up ↗

down ↘

level →

Verb

climb (up) ↗

go up ↗

increase ↗

rise ↗

decline ↘

decrease ↘

dip ↘

drop ↘

fall ↘

go down ↘

Level

it doesn't change →

it doesn't move →

it remains constant →

Sales have

climbed ↗

gone up ↗

increased ↗

risen ↗

Production has

declined ↘

decreased ↘

dipped ↘

fallen ↘

Typical nouns: Manufacturing, sales, productions etc.

The increase in

The rise in

The decline in

The dip in

The fall in

Appendix 3: Technical Jobs - What

Name	What does he do?	German
crane operator	drives a crane	Kranführer
air traffic control	gives airplanes their route into and out of airports	Fluglotse
apprentice	is trained to acquire the skills necessary for a job	Auszubildende(r)
architect	designs the layout of houses	Architekt
assistant to the Board	carries out routine support for the Board	Vorstandsassistent
automotive engineer	designs vehicles	Fahrzeugingenieur
blue-collar worker	works with his hands mostly in production	einfacher Arbeiter
carpenter	makes things from wood, e.g. cupboards, drawers	Schreiner
carpet-layer	lays the carpet	Teppichverleger
chemical engineer	designs fertilisers, drugs, detergents	Chemiker
civil engineer	designs bridges, roads etc.	Bauingenieur
clerk	does routine office work, white-collar worker	Bürokraft
computer operator	knows all about word processors, spread sheets	Bildschirmarbeiter; EDV...
contractor	Supplies customers with materials	Bauunternehmer; Zulieferant
decorator	redecorates your flat, puts up new wallpaper	Innenausstatter, Tapezierer
department leader/ head of department	is responsible for work and workers in an office	Abteilungsleiter
design engineer	designs products	Entwickler; Konstrukteur
designer	he creates the style of products	Designer, Stilist
director	is responsible for a whole division or company	Direktor
draughtsman	creates the drawings with which the engineer works	technischer Zeichner

electrician	helps in case of a black-out	Elektriker
executive	top manager	Führungskraft
fitter	installs and mounts mechanical parts	Monteur
foreman	is responsible for a team of workers	Vorarbeiter
glass-blower	forms liquid glass into shapes	Glasbläser
glazier	cuts and fits panes of glass	Glaser
graphic arts designer	designs layouts of printed materials	Graphiker
interior decorator	designs the inside of a house	Innendekorateur
joiner	makes beams and floors etc. for houses	Zimmermann
lab assistant	does support work in the lab	Laborassistent
lab technician	assists in some way in the lab	Labortechniker
locksmith	makes keys and other small metal objects	Schlosser
maintenance fitter	he maintains engines from time to time	Wartungsmechaniker
manager	supervises clerks etc.	Manager
mason	builds houses, lays bricks	Maurer
mechanic	repairs cars and engines	Mechaniker
navigator	guides planes and plots the route	Navigator
optician	repairs your spectacles	Optiker
painter	paints the walls in your house	Maler
papermaker	makes paper from wood	Papiermacher
part-time worker	works only a few hours a day	Teilzeitbeschäftigter
physicist	knows why an apple drops to the ground	Physiker
pilot	flies planes	Pilot
plasterer	plasters/smooths walls	Gipser; Stukkateur
plumber	repairs toilets and pipes	Installateur

printing press operator	operates a printing press	Drucker
production engineer	guides and plans production and manufacturing	Fertigungsingenieur
programmer	writes programs that tell the computer what to do	Programmierer
research engineer	develops new products	Entwicklungsingenieur
roofer	puts tiles on the roof	Dachdecker
secretary to the director	completes routine tasks for a boss	Chefsekretärin
semi-skilled worker	has been trained to do one or two jobs efficiently	angelernter Arbeiter
skilled worker	has learned the skills of the trade	Facharbeiter
smith(y)	forges/shapes hot metals by hammering	Schmied
software engineer	designs and creates software	Softwareentwickler
systems analyst	builds up processes and systems	Systemanalytiker
technician	assists with a few more simple technical jobs	Techniker
temporary worker	works for a short time and leaves the company	Gelegenheitsarbeiter
tiler	puts tiles on the walls in bathrooms/kitchens	Fliesenleger
toolmaker	makes the metal tools which do other jobs	Werkzeugmacher
unemployed; jobless	is looking for a job	Arbeitslose(r)
unskilled worker	has not done an apprenticeship	ungelernter Arbeiter
welder	joins two pieces of metal together with heat	Schweißer
white-collar worker	works in an office at a desk	Angestellter

Appendix 4: Comparing weights and measures

What do we do when we are confronted with weights and measures from other countries? It's a bit frustrating but with the help of a pocket calculator you can manage it:

British/American weights and measures

Length

1 inch	=25.4 millimetres	
1 foot	= 12 inches	= 0.3048 metres
3 feet	= 1 yard	= 0.9144 metres
1,760 yards	= 1 mile	= 1.609 kilometres

Area

1 square inch	= 6.452 square centimetres	
144 square inches	= 1 square foot	= 9.29 square decimetres
1 square yard	= 9 square feet	= 9.836 square metres
1 acre	= 4840 square yards	= 0.405 hectares
1 square mile	= 640 acres	= 259 hectares

Weight

1 ounce	= 28.35 grams	
1 pound	= 16 ounces	= 9.4536 kilograms
14 pounds	= 1 stone	= 6.35 kilograms

Great Britain

1 hundred weight (cwt)	= 112 pounds	= 50.80 kilograms
1 long ton	= 2,240 pounds	= 1,016.05 kilograms

USA

100 pounds	= 1 hundredweight (cwt)	= 45.36 kilograms
2,000 pounds	= 1 short ton	= 907.18 kilograms

Liquids:

Great Britain:

1 pint = 0.568 litres
2 pints = 1 quart
4 quarts = 1 imperial gallon = 4.546 litres

USA:

1 pint = 0.4732 litres
1 gallon = 3.7853 litres

Temperature

Celsius	Fahrenheit
-17.8°	0°
-1°	14°
0°	32°
10°	50°
20°	68°
30°	86°
40°	104°
50°	122°
etc.	etc.

To convert Celsius into Fahrenheit: multiply by 9, divide by 5 and add 32.
To convert Fahrenheit into Celsius: subtract 32, multiply by 5 and divide by 9.

Bildquellen

Überflieger

Die «Überflieger» sind der Einstieg für alle, denen ein ganzes Lehrbuch zu langwierig und ein Sprachführer zu wörterbuchhaft ist. Schon in wenigen Tagen kann man damit die notwendigen Grundkenntnisse erwerben, um sich in einem fremden Land zu verständigen. Damit Urlaub und Geschäftsreise nicht nur sprachlich ein voller Erfolg werden, gibt es außerdem praktische Tips zu Kultur und Alltag.
Eine Auswahl der lieferbaren Titel:

Uwe Kreisel / Pamela Ann Tabbert
American Slang in letzter Minute
(Buch: rororo 19623, Buch mit Cassette: rororo 19624, Cassette: rororo 19705)

Hanne Schönig / Hatem Lahmar
Arabisch in letzter Minute
(Buch: rororo 19541, Buch mit Cassette: rororo 19542, Cassette: rororo 19700)

Petra Schaeber
Bralilianisch in letzter Minute
(Buch: rororo 19977, Buch mit Cassette: rororo 19979, Cassette: 19978)

Isabelle Jue / Nicole Zimmermann
Französisch in letzter Minute
(Buch: rororo 19628, Buch mit Cassette: rororo 19629, Cassette: rororo 19702)

Frida Bordon / Giuseppe Siciliano
Italienisch in letzter Minute
(Buch: rororo 19626, Buch mit Cassette: rororo 19627, Cassette: rororo 19703)

Elisabeth Völpel
Portugiesisch in letzter Minute
(Buch: rororo 19686, Buch mit Cassette: rororo 19687, Cassette: rororo 19736)

Dorothee Bernhardt
Russisch in letzter Minute
(Buch: rororo 19797, Buch mit Cassette: rororo 19799, Cassette: rororo 19798)

Christof Kehr
Spanisch in letzter Minute
(Buch: rororo 19526, Buch mit Cassette: rororo 19527, Cassette: rororo 19701)

Karl-Heinz Scheffler
Türkisch in letzter Minute
(Buch: rororo 19688, Buch mit Cassette: rororo 19689, Cassette: rororo 19735)

rororo sprachen

rororo sprachen wird herausgegeben von Ludwig Moos. Ein Gesamtverzeichnis der Reihe finden Sie in der *Rowohlt Revue*. Vierteljährlich neu. Kostenlos in Ihrer Buchhandlung.